MIMESIS
INTERNATIONAL

OUT OF SERIES

BESTIARIUM

Human and Animal Representations

Edited by
Mariaelisa Dimino, Alessia Polatti, Roberta Zanoni

MIMESIS
INTERNATIONAL

This book has been published with the financial subsidy of the Department of Human Sciences, University of Verona.

© 2018 – MIMESIS INTERNATIONAL
www.mimesisinternational.com
e-mail: info@mimesisinternational.com

Isbn: 9788869771248
Book series: *Out of series*

© MIM Edizioni Srl
P.I. C.F. 02419370305

TABLE OF CONTENTS

MARIAELISA DIMINO, ALESSIA POLATTI, ROBERTA ZANONI
INTRODUCTION

In the last decades, new achievements in fields such as Ecology and Cognitive Ethology have created the social need to deeply reconsider the ethical status of animals. From a theoretical point of view, these peculiar social demands have imposed an interpretative shift in the Humanities, leading to the so-called 'Animal Turn' in cultural studies (Harriet Ritvo, 'On the Animal Turn', 2007).

Indeed, if on the one hand the contribution of scholars such as Jacques Derrida (*L'Animal que donc je suis*, 2006), Giorgio Agamben (*L'Aperto: l'uomo e l'animale*, 2002), Cora Diamond (*The Realistic Spirit*, 1991), and J. M. Coetzee (*The Lives of Animals*, 1999) has allowed to dismiss the conception according to which 'animals were mere blank pages onto which human wrote meaning' (Erica Fudge, 'The History of Animals', 2006), on the other hand, it has demonstrated the difficulty of abandoning the anthropocentric point of view which has always characterized human discourses on animals. This theoretical turn also raised some fundamental questions about Otherness, human-animal relationships, the ontological status of animals, and the meaning of humanity and animality. As a result, the traditional epistemological categories of the Humanities have been called into question. What are the ontological, phenomenological, and ethical differences emerging from the comparison between human and animal? How does the distinction between humanity and animality change over time and in different cultural contexts? How does the interpretation of a text change when trying to assume a non-anthropocentric point of view on the representation of the animal? Which linguistics strategies are deployed when speaking of animals and what do they reveal?

In order to discuss these and other questions, in 2016 we organised an international meeting at the PhD School of Humanities of the University of Verona. Our project aimed at gathering PhD students and young researchers coming from prestigious universities from all over the world, in order to support a constructive debate on the current theories on the human-animal relations in the Humanities. The result was a lively discussion focused on texts and discursive practices which revealed the epistemological and cultural dynamics structuring the contemporary representation of the animal.

The outcomes of the conference's discussion are collected in the present volume which, as a sort of contemporary bestiarium, presents and describes a gallery of 'animals' demonstrating the developments brought about by a new awareness on the subject in the field of the Humanities. The animals portrayed in this book thus appear in a wide variety of forms, ranging from literary representations, to philosophical and artistic depictions.

Hence the essays collected in the volume are principally focused on Literary and Cultural Animal Studies. On the one hand, as Roland Borgards points out, 'literary animal studies provide an original contribution to the overarching aim associated with the animal turn, as it explores literary animals in a way that highlights how deeply aesthetic and literary questions are interwoven with both political and material spheres'.[1] On the other hand, the development of CAS provides a new perspective on human-animal relationships, above all in philosophy and art.

The philosophical debate on the issue of the animal has extended, from the traditional fields of ethics and ontology, also to other branches such as political philosophy. It is in the arts, however, that the perspective of animal studies has revealed all its potential, demonstrating that the questioning of the traditional assumptions on the ontological difference between animals and humans implies first of all a revision of our own stance and practises as humans.[2]

Indeed, at the intersection of all the animal portrayals taken into account, which cross cultures and our everyday life, the essays collected in this book reflect on the stance of the animal from various theoretical perspectives, which mirror the contemporary debates and considerations affecting contemporary practises and discourses. The contributors alternatively endeavour to confront the most recent perspectives of Animal Studies in their fields of research, trying to adopt a critical attitude capable of shortening the ontological distance between the human and

---

1   Roland Borgards, 'Introduction: Cultural and Literary Animal Studies', in *JLT* (2015); 9(2): De Gruyter, 155-160, (p. 156).
2   Roland Borgards, 'Einleitung: Cultural Animal Studies', in Roland Borgards (Hrsg.), *Tiere. Ein kulturwissenschaftliches Handbuch*. Stuttgart, (2016), 1-6, (p. 4).

the animal. In this questioning of the anthropocentric 'belief in absolute human uniqueness',[3] the body becomes the joining link between the animal and the human, in Agamben's words a 'zone of indifference [...] within which [... t]he articulation between human and animal, man and non-man, speaking being and living being, must take place'.[4]

The essays by Lucia Zaietta, Richard Hutchins, Amadeusz Just, Benedetta Piazzesi, and Peter Kofler share an interest in the body and in the discourses on human and animal bodies, variously declined, particularly in relation with literature, philosophy, and art. The arts, indeed, have given a strong contribution to the contemporary conception and approach to the body in Animal Studies. The essays show an approach which 'still retain[s] the power to rekindle that deep time when the boundary between human and animal was permeable, when humans knew they were one among many other animals.'[5] The essays share a Critical Animal Studies approach which 'provide us the means of deepening our demand for animal ethics [...] by interrogating rather than assuming the categories of 'human' and 'animal.'"[6]

The theories of Merleau Ponty, highly influential in the contemporary Critical Animal Studies debate, are dealt with by Lucia Zaietta in her essay 'An Incorporated Meaning. Corporeity, Visibility and Symbolism in the Animal World.' The author affirms the significance of corporeity as an attribute of all living beings, humans and animals, thus equated by their being a body. The body – both the animal's and the human's one – becomes, in Merleau Ponty's reflection, expressive and communicative, only due to its being in the world. Behaviours and gestures open the bodies to intersubjectivity, they represent intentional actions performed by the body thus allowing it to interact with the environment surrounding it. Each animal's behaviours and gestures give rise to an animal semiotics: the presence of the animal is significant in itself. In her essay Zaietta analyses intersubjectivity among animals,

---

3   Ralph Acampora, *Corporal Compassion: Animal Ethics and Philosophy of Body*. (Pittsburgh: University of Pittsburgh Press, 2006), p. 22.

4   Giorgio Agamben, *The Open: Man and Animal*, trans. by Kevin Attell, (Stanford: Stanford University Press, 2004), 37, 38.

5   Marion W. Copeland, 'Literary Animal Studies', in *Where We Are, Where We Are Going*: 2012, 91-105, Abstract.

6   Kimberly W. Benston, 'Experimenting at the Threshold: Sacrifice, Anthropomorphism, and the Aims of (Critical) Animal Studies. Publications of the Modern Language Association', *PMLA*, Volume 124, Number 2, March, 2009, 548-555, (p. 553).

humans, and between animals and humans. She illustrates how the relations between the subjects as bodies pass through intercorporeity, a communication through the meaningful presence of the body in the world, and intervisibility, a result of the visual scenarios created in the process of performance of the instinctual behaviours. The human becomes, in this reflection, one among other animals.

The body in its functioning links the animal and the human: the similarity of the animal ontology with the human one goes against the anthropocentric belief in human uniqueness. The relationship between the animal and the human body and their inter-exchangeability are investigated also in the essay by Richard Hutchins 'Making the Absent Referent Present: the Sacrifice of Women and Animals in Lucretius.' The concept of the absent referent – indicating to the words used in order to conceal the true animal referent of the meat humans are eating – is declined in Hutchins' essay from an eco-feminist perspective. The woman and the animal become each other's absent referent in the male power discourses. Following Carol Adams' theory Hutchins connects the abuse of animals with the abuse of women in a patriarchal society. The author takes two texts of Lucretius' epic poem, *De rerum natura* (or *On the Nature of Things*) as examples. The text regarding the sacrifice of Iphigenia depicts the woman as an animal while the one describing a mother cow searching for her calf sacrificed on an altar connotes the animal in human terms. Through Adam's reading by Hutchins, thus, the animals are the absent referent when people eat meat, and they also become the absent referent in descriptions of women's suffering. The two stories are, thus, related and need to be read in conjunction in order to appreciate the duplication of the absent referent and the social critique put forward by the author. The reflection conducted in the essay tries to equate the human and animal suffering under a prevalently male rule.

Hutchins' essay moreover, by describing the mother cow as a human mother, equates their suffering contributing to the philosophical debate on animal suffering which is thoroughly analysed by Amadeusz Just in his essay 'Schopenhauer and Diamond on Animal Ethics.' Following the theories of Shopenhauer and Diamond the author dismantles the claims that cruelty to animals may lead to cruelty to people and the *arguments from common capacities*. Both arguments, although convincing for some people and thus generating a positive outcome, are not considered valid enough and genuine from a philosophical point of view, being easily dismantled. The philosophical reflection of Just considers the debate regarding the eligibility of the animal as an ethical subject. Animals are seen by the two philosophers as participating in morality with human beings. The argumentation of Just tries to provide a perspective on the human-animal relation based on the observation of our current relationship with animals and on the idea of our capacity of treating them with compassion just depending on our morals and not on external promises of prizes or punishments.

The ethical question developed in philosophy inevitably influences also other fields of research and, in particular, the scientific one in which the animal occupies a controversial position. Body and animality are analysed from a scientific perspective in Benedetta Piazzesi's 'Scientific Bestiarium: the Living, the Dead, and the Normal,' in which the author takes into consideration the different representations of the animal body by scientific literature over the years. Piazzesi's excursus starts from Aristotle's contribution to science in his analysis of corpses. The body of the animal becomes, for Aristotle, a source of knowledge allowing to classify the species and to compare them, according to an approach that will be resumed by the moderns. Piazzesi's reflection focuses on the works by Francesco Redi and Louis Pasteur, the forefathers of parasitology and epidemiology. The above-mentioned studies often involved the starvation of the animals, observed as a common practice which raised no moral issues in the scientists. In order to obtain scientific results, the animal body becomes an object of research, infinitely repeatable and expendable. The individuality of the animal, however, cannot completely be neglected: the author wishes for a deeper investigation on the history of zootechnics and for a clearer critical approach in order to answer the questions regarding scientific research and its interaction with the government of human and animal life.

The body of the animal figuratively 'phagocytised' by the human is the subject of Peter Kofler's 'The Anachronous Montage of Man and Serpent.' The author's reflection starts from the experiences and thoughts contained in the *Lecture on Serpent Ritual* by Aby Warburg. He recalls Warburg's description of the rites of some Native American tribes through which the humans try to transform themselves into animals to acquire their power or to anticipate the actions of the hunt; the dance ends with the totemic union of the man with the animal. The same juxtaposition is observed by Kofler in art, in particular in the *Laocoön and his Sons* statue. The fight of the man with the serpent is represented in the statue as a continuum of bodies and of feelings which are physically and symbolically intertwined. Kofler, then, goes on analysing a 13th Century fresco found on the lower part of the apse of St. Jakob in Kastelaz (BZ) representing hybrid figures. The fresco shows half human half animal figures, holding animals in their hands or phagocytising them. The assimilation with the animal is carried out in order to gain its powers but also in order to underline the animals' identification with the human subconscious, unknown, and foreign. The relationship between the human and the animal appears, in some cases, mutual because of the complex entanglement of the bodies recalling *Laocoön*'s one. Warburg's *Lecture on Serpent Ritual* shows that the aforementioned representations of the serpent – the Moki dance, the *Laocoön*, and the fresco – testify to its double nature, compared to that of the Derridean *pharmakon*, both poisonous and endowed with healing powers. The description of the serpent, moreover, affects the writing of Warburg, which acquires the ambivalent traits of the subject he is describing, in a sort of hybridization.

Examples of human/animal hybridity are traced also in the essays dealing with Literary Animal Studies. In these papers, hybridity is linked to questions of language, writing, and metamorphoses which should give rise to a human/animal equal right to self-expression. In this light hybridity acquires the meaning of a veritable transformation, a literary and cultural vehicle used to understand the human/animal paradigm which 'functions in something like the way that caricature functions, by greatly exaggerating a given feature as a way of focusing attention on it'.[7]

An example of this condition is given by the 'Re Porco' of Straparola's fairy-tales analysed in Flavia Palma's essay 'Changing Shapes: Human and Animal Metamorphoses in Straparola'. Palma's work sheds new light on the role of animals in fairy-tales, a literary genre in which the role of animality is often overshadowed in favour of an anthropocentrism typical of the Renaissance thinking. Palma's essay, however, traces a first attempt to overcome this condition. The author actually demonstrates that the assumption according to which men are superior to animals by God's will – that is a perfect representation of the Renaissance standpoint – is challenged for the first time in the mid- and late sixteenth century by Straparola's fairy-tales. These peculiar stories do not make a strong distinction between animals and men, but depicts a man who goes beyond the limits imposed by his human nature in order to survive and achieve his goals. From this perspective, the

---

7   Harvey L. Hix, 'Hybridity Is the New Metamorphosis', in *Comparative Critical Studies* 9.3, (Edinburgh University Press: 2012), 271-283, (p. 271).

metamorphosis of Straparola's heroes into a rational animal assumes a fundamental role for the achievement of a happy ending and in order to underline the distrust not on human abilities, but on their omnipotence. In other words, Palma seeks to demonstrate that, in Straparola's tales, 'human beings are not considered as powerful as they were in the previous decades, so that they need to resort to something which is non-human or extra-human in order to have some chance to succeed'. Thus, Straparola's protagonists are far from the Machiavellian image of the prince: they are instead a notable representation of the growing distrust in the supremacy of men's superiority.

If on the one hand the power of words is not so relevant in Straparola's tales, since the pig-prince cannot talk, on the other hand the relevance of language is well depicted in Simone Rebora's essay on James Joyce's *Finnegans Wake*. In particular, in his "It's as Semper as Oxhousehumper!' James Joyce's Animalization of the Human', Rebora highlights the linguistic hybridity of Joyce's work, through which animals can show all the limits and the contradictions of human nature: as a result, human superiority is introduced as a delusion to the reader. In this context, the author analyses both Joyce's depiction of some animals' figures employed to explain the origin of language, knowledge, and time, and a series of hybrid human/animal creatures which incarnate the salvation of humanity in reaction to the horrors brought about by technological evolution and sustained by the affirmation of psychanalysis. What is more, this particular kind of hybridization can be associated to Kristeva's idea of 'liquidation, liquefaction, of both feminine and masculine, in the flow of a style, halting at no one identity – whether personal, ideological, or sexual – but knowing them all'[8] which, according to Rebora, can be seen also as a liquefaction of humanity and animality in Joyce's works.

The relationship between hybridity and human/animal identity is also at the core of Eirini Apanomeritaki's essay 'The Animal as Writer: Creativity and Melancholia in Marie Darrieussecq's Pig Tales'. Through an interesting investigation of Darrieussecq's novel *Pig Tales*, whose protagonist gradually becomes a pig writing her own story, Apanomeritaki reads the animal/human subject as melancholic and

---

8   Julia Kristeva, 'Joyce 'The Gracehoper' or the Return of Orpheus', trans. by Louise Burchill, rev. by Jacques Aubert and Shari Benstock, in *James Joyce: The Augmented Ninth*, ed. by Bernard Benstock (Syracuse: Syracuse University Press, 1988), 167-180, (p. 179-180).

abject. This is possible because the pig's ability to write is prompted by a loss, that of her wolf mate: in this way, not only does the author suggest that animals are in a certain sense better than humans since they can live peacefully together – considering the pig-wolf's love affair –, but also that the relationship between a metamorphosed, non-metaphorical, subject and the writing creativity is fully expressed through the mourning. Moreover, the urge to write and, as a consequence, to establish and communicate a sense of identity, is encouraged by the protagonists' inexplicable metamorphosis which transforms her identity into a veritable hybrid feature. Starting from these premises, Apanomeritaki shows that the act of literary writing can directly emerge only after melancholia, thus differentiating between human communication and a story which comes from a metamorphosed human-animal being.

The need of overcoming the traditional tendency of literary criticism to read the animal merely as a symbol, a metaphor or an allegory, whose only purpose is that of representing and negotiating human power relations of race, class, and gender, has been variously signalled in Literary Animal Studies. As Marie Cazaban Mazerolles observes in her essay 'St. Mawr, Lassie and a 'Naive Hedgehog': Theoretical and Practical Issue about Representing the Animal *qua* Animal in Literary Narratives', 'over the last century, a growing number of writers in line with the global reassessment of animal nature and our human relationships with them have opposed such a use of animal figures, and ostensibly strived to reach a non-allegorical, non-symbolical representation of the animal as an animal'. And yet, as Cazaban points out, the aim of overcoming an anthropocentric perspective in LAS is still far to be reached. In her essay, the author examines three texts, D.H. Lawrence's short story 'St. Mawr' (1925), Éric Chevillard's 'Du hérisson' (2002), and T.C. Boyle's 'Heart of a champion' (1974), which, according to the view of contemporary animal studies discourse, supposedly deliver a 'non-allegorical, non-symbolical representation of the animal as an animal.' As Cazaban demonstrates, although apparently overcoming traditional models of animal representation, these texts actually are affected by three fundamental anthropological biases: the Platonic bias, i.e. the tendency to talk about the animal in essentialist terms, as a homogeneous category or idea; the Aristotelian bias, i.e. treating the animal's individuality as a synecdoche for its species; and the Pygmalion bias, i.e. subsuming animal subjectivity under a human perspective. The ultimate elusiveness of the animal figure within human literary representation, its being 'un être de fuite', as Cazaban quotes, seems to suggest that the only 'genuinely successful way out of anthropocentrism ultimately consists in the detachment and remoteness of the animal figure', at the same time pointing at the need for the field of LAS to constantly put under scrutiny its own methodology and theories. Indeed, in spite of the difficulty of overcoming an anthropocentric perspective from an inescapable human point of view, an attempt can be made, at least, at deconstructing traditional assumptions on the animal as something deprived of agency, that is, speechless and passive. This is the way paved by Derrida (2002), whose arguing for the need to meet the animal gaze is also the starting point of Valentina Savietto's analysis of the dynamics of the gaze in Botho Strauß's works *Der junge Mann* (1984) and *Schlußchor* (1991).

Following the different manifestations of animals and animality in Strauß's works, which range from mythical constructions to ritual cannibalism, from predatory behaviour to the problem of inhabiting the same environment with man, Savietto's essay 'Botho Strauß: Mythology and Symbology within the Conception of a 'Human Animality'' shows how the author's insistent linguistic ambiguity ends up in a deferment of meaning which destroys the assumed borders between the human and the animal. Quoting Wolfe's statement, that 'the animal, when you think about it, is everywhere'[9] Savietto projects Strauß's questioning of ontological, identity and gender divisions on the space of a 'humanimality', which allows for individuals and groups to overcome traditional representations of the human and the animal in socially constructed norms and discourses.

The question of socially constructed norms and the boundaries of civilization are dealt with by Judith Rahn's 'Human Beasts: Eating Meat as a Negotiation of Self' as well. Rahn starts from Freud's interest on cannibalism developed in *Totem and Taboo*, according to which this act is posited as 'the dark secret behind the origins of civilization'. This notion sees cannibalism in-between the ideas of 'civilized' and 'uncivilized', as well as in-between the notions of culture and of a natural and wild animality. Following Rahn's argumentation, the cannibal is at the cross-roads between consumption of and kinship with animals, so that cannibalism can be described as an extreme and pure form of hybridity, a state of in-betweenness, where the Other is the non-self but it also has the potential to turn into a part of the self by means of incorporation.

---

9   Cary Wolfe, 'Human, All Too Human: 'Animal Studies' and the Humanities', *PMLA*, Vol. 124, No. 2 (Mar., 2009), 564-575, (p. 564).

Cannibalism, therefore, encourages a strangely intimate affinity between the eaters and what is eaten, and creates an unavoidable relation with the Other. The author links this theoretical background to Margaret Atwood's recent trilogy *MaddAddam*. In the Canadian writer's novels, 'questions of consumptions are negotiated by means of synthetically breed human-animal hybrids' which constitute Margaret Atwood's ironic vision of the human future, along with a newly created human race that rejects all animal food in favour of plant-consumption. The essay then insists on the correlation of animal meat consumption and anthropophagy as a necessary force in a sort of reconstruction and renegotiation of the human, the animal and the edible in a futuristic society.

The questioning of a socially constructed representation of the animal is also at the core of Elena Ogliari's contribution 'Humans as Zoo Antelopes and White Swans: Animal Imagery in Henry James's 'Julia Bride'', which examines the use of animal imagery in Henry James's 'Julia Bride'. Starting from John Berger's claim of the disappearance of the real animal in capitalistic societies, Ogliari discusses how modern transformation of the animal into pure spectacle for economical purposes – epitomised in the social institution of public zoos – is projected onto the figure of Julia, James' female protagonist. This demonstrates how the logic of early twentieth-century financial capitalism ends up marginalising both women and animals. Moreover, by drawing a parallel between the animals on display in the Central Park Zoo or in the London Zoo visited by James in 1869 and Julia, who 'cannot absent herself from the others' gaze', Ogliari demonstrates how 'James is not adamant about the separation between animal and humans, because even the latter may be treated as entertaining living exhibits: Julia and her mother resemble animals in cages as all their acts are being examined and commented upon'.

Reversing the perspective on this act of 'zoological' observation, the last contribution in this 'bestiarium' reports the experience of two German artists, Daniela Maria Hirsch and David Gaehtgens, whose video installation was on display during our eponymous international conference. The two artists have gathered video recordings of the visitors of the Berlin Zoo looking at the animals on display. Starting from the assumption that '[t]he setting of the zoo focuses prototypically on central aspects of the human posture towards non-human animals, but also on human organization itself', Gaethgens and Hirsch have directed their cameras at human animals in order to show how these latters' attitude towards co-animals is affected by the organization of space and is oriented by an 'overall concept of management and application for use'. The main interest of the two artists is indeed 'observing the observer', a practice which involves excluding from the main frame the actual object of the act of observation, i.e. the animal, in order to convey all attention to human animals which are seen 'both as an overall appearance and in the individuals' details'. By the means of 'direct cinema', 'longterm recordings', and 'participant observations', Gaethgens and Hirsch have collected the results of their own observations in a corpus 'of visual knowledge', which has eventually transformed their own 'perception of the zoo into an allegory for the human civilization project'. At the same time, their experiment also shows all its productive potential in installative situations, as the two artists remark a 'continuous re-search: to look and look again and look again'. This process may lead human animals to accept the invitation to meet others' and their own animal gaze.

We would like to thank the peer-reviewers Luigi D'Agnone, Francesca Dainese, Giulia Pellegrino, Marco Robecchi for their contribution in reading and selecting the essays.

Sincere thanks also to the PhD School of Humanities and the PhD Programme in Modern Languages, Literatures, and Cultures of the University of Verona for supporting us in all respects and making this volume possible.

Bibliography

Acampora Ralph, *Corporal Compassion: Animal Ethics and Philosophy of Body*. (Pittsburgh: University of Pittsburgh Press, 2006).

Agamben Giorgio, *The Open: Man and Animal*, trans. by Kevin Attell, (Stanford: Stanford University Press, 2004), 37, 38.

Benston Kimberley, 'Experimenting at the Threshold: Sacrifice, Anthropomorphism, and the Aims of (Critical) Animal Studies. Publications of the Modern Language Association', *PMLA*, Volume 124, Number 2, March, 2009, pp. 548-555.

Borgards Roland, 'Einleitung: Cultural Animal Studies', Roland Borgards (Hrsg.), *Tiere. Ein kulturwissenschaftliches Handbuch*, (Stuttgart: J. B. Metzler, 2016), pp. 1-6.

Borgards Roland, 'Introduction: Cultural and Literary Animal Studies', in *JLT* (2015); 9(2): De Gruyter, pp. 155-160.

Coetzee J. M., *The lives of Animals*, (Princetone: Princetone University Press, 1999).

Copeland Marion W., 'Literary Animal Studies', *Where We Are, Where We Are Going*, 2012, pp. 91-105, Abstract.

Derrida Jacques, *L'Animal que donc je suis*, (Paris: Galilée, 2006).

Diamond Cora, *The Realistic Spirit*, (Cambridge: The MIT Press, 1991).

Fudge Erica, 'The History of Animals', *HNet*, (2006). https://networks.h-net.org/node/16560/pages/32226/history-animals-erica-fudge

Hix Harvey, 'Hybridity Is the New Metamorphosis', *Comparative Critical Studies* 9.3, (Edinburgh University Press: 2012), pp. 271-283.

Kristeva Julia, 'Joyce 'The Gracehoper' or the Return of Orpheus', trans. by Louise Burchill, rev. by Jacques Aubert and Shari Benstock, *James Joyce: The Augmented Ninth*, ed. by Bernard Benstock (Syracuse: Syracuse University Press, 1988), pp. 167-180.

Ritvo Harriet, 'On the Animal Turn', *Daedalus*, Volume 136, Issue 4, Fall 2007, pp. 118-122.

Wolfe Cary, 'Human, All Too Human: 'Animal Studies' and the Humanities', *PMLA*, Vol. 124, No. 2 (Mar., 2009), pp. 564-575.

LUCIA ZAIETTA

AN INCORPORATED MEANING.
CORPOREITY, VISIBILITY AND SYMBOLISM
IN THE ANIMAL WORLD

Introduction

Maurice Merleau-Ponty's phenomenology offers an original approach to the reconsideration of animality and the relationship between humans and animals. Through a reformulation of classical notions, Merleau-Ponty addresses his research to reviewing the being-animal question and identifying a symbolism inherent in animal conduct. In this article, I focus on Merleau-Ponty's course notes about nature, particularly on his interpretation of Adolf Portmann and Konrad Lorenz. Portmann's investigation on animal forms reveals the richness of animal appearance: animals are subjects of their own corporeity and their own behaviour, as well as possessing completely 'useless' and aesthetic expressive values which exist to be seen. This intrinsic visibility proves the immanent relationality among animals (what Merleau-Ponty calls *Inter-animality*) and within nature (what Merleau-Ponty calls *Inter-weaving*). Similarly, in Konrad Lorenz's analysis of instinct, we find an empty activity that does not follow the law of utility. Far from being a reductive sum of stimulus-response, instinctual behaviour reveals purposeless conduct. These narcissistic modalities allow for an original comprehension of animality and reveal the symbolism, as well as the pre-culture, immanent in corporeity itself. According to Merleau-Ponty, corporeity becomes the key to an authentic understanding of animality and the emergence of the *human architectonic*. In this way, each living being is a *manner of being a body* and takes part in a meaningful horizon that is the Being itself.

Towards an Animal Stylistics

In Merleau-Ponty's phenomenology, corporeity is a central notion and a main issue. Throughout his philosophy Merleau-Ponty conceives the body as the core of perception and movement, as well as the primary way of being-in-the-world. Moreover, the body is what animals and human beings share: both possessing a primordial access to the world that is the body, the point zero of any movement and perception.

Animality is a central point in his first work, *The Structure of Behaviour*, and the analyses Merleau-Ponty conducts are based on the corporeity of the organism. It is no coincidence that he chooses the term 'organism', that is, a neutral term, to reflect on the significance of the living being. Here, Merleau-Ponty is deeply influenced by the holistic approach of Kurt Goldstein. The atomistic conception, according to which an organism is nothing but an assemblage of parts juxtaposed in space, does not describe the essence of the organism, its meaningful organization. There is something that eludes this reductive comprehension. Far from reducing the animal to a machine, Merleau-Ponty stresses the immanent signification of each living being. The *true* organism is a concrete, significant, and indecomposable totality. In this view, behaviour is not a mere response to a stimulus but it is a significant reaction to a signal that the organism is able to recognize. In these analyses, there is no longer a classification based on a hierarchy from the simplest to the most complex, but a phenomenological description of behaviour. Merleau-Ponty is looking for a different way to conceive the meaning of a living being and its behaviour: what emerges is neither a mechanistic conception, nor an intellectual one. An organism's conduct has a sense, a directive activity, which cannot be explained by classical approaches. Behaviour is not a blind mechanism, nor some sort of intelligence descended into the body; rather it is an oriented activity. According to Merleau-Ponty:

> The subject does not live in a world of states of consciousness or representations from which he would believe himself able to act on and would know external things by a sort of miracle. He lives in a universe of experience, in a milieu, which is neutral with regard to the substantial distinctions between the organism, thought and extension; he lives in direct commerce with beings, things and his own body.[1]

---

1   Maurice Merleau- Ponty *The Structure of Behaviour*, trans. by Alden L. Fisher (Boston: Beacon Press, 1967), p. 189.

Behaviour is a meaningful activity that presupposes something more than a merely physical event but, at the same time, does not necessarily encompass an intelligent dimension. In this way, Merleau-Ponty reformulates the definitions of organism, behaviour, and finally, the notion of subjectivity:

> From this moment on behaviour is detached from the order of the in-itself (*en soi*) and becomes the projection outside the organism of a possibility which is internal to it. The world, inasmuch as it harbors living beings, ceases to be a material plenum consisting of juxtaposed parts; it opens up at the place where behaviour appears.[2]

Authentically understanding behaviour means, first of all, recognizing that there is a being-for-the-animal, a specific way of being in the world. The starting point of the investigation is the organism (a neutral term that includes animals and human beings) defined by a significant openness to the world, as a unique way of living and behaving. It is what Merleau-Ponty calls être-*au-monde*: a spatial subjectivity, a situated subject that is not yet a full consciousness, and no longer a mechanical system. It is a significant corporeity. In this way, we arrive to the central notion of our reasoning, which is the body. As everybody knows, corporeity is the heart of the phenomenology of Merleau-Ponty, but it is also fundamental in our reconsideration regarding animality. Animals and human beings share a corporeity, which is the seat of their openness to the world. Corporeity opens to a structural ambiguity, the body being interpreted neither as a pure subject nor as a pure object but as a natural subject. The body, in Merleau-Ponty's account, is the 'absolute here', the installation of the first and original coordinates. And it is in the body itself that we can find a primordial and original symbolism. According to Merleau-Ponty, every living being *is* its own body. The body, including that of the animal, is a total experience, the reason why every organism is open to the world. In *Phenomenology of Perception*, the body becomes the central notion in order to comprehend subjectivity and radically rethink phenomenology. According to Merleau-Ponty:

> If the subject is in a situation, or even if the subject is nothing other than a possibility of situations, this is because he only achieves his ipseity by actually being a body and by entering into the world through

---

2   *The Structure of Behaviour*, p. 167.

this body. If I find, while reflecting upon the essence of the body, that it is tied to the essence of the world, this is because my existence as subjectivity is identical with my existence as a body and with the existence of the world, and because, ultimately, the subject that I am, understood concretely, is inseparable from this particular body and from this particular world. The ontological world and body that we uncover at the core of the subject are not the world and the body as ideas; rather, they are the world itself condensed into a comprehensive hold and the body itself as a knowing-body.[3]

The body 'takes root' in the situation: there is a situation because there is a body. The body opens up to the possible and virtual dimension. But there is something more, because the body *expresses something* and it communicates co-belonging [*co-appartenance*] and intertwining among living beings within nature. The body has an expressive value, it is expression and communication. Through gestures, through attitude, through presence itself, a body means *something*. As we have seen, in Merleau-Ponty's perspective corporeity is the correlational structure, an opening towards the outside and towards the other. It is through the expressivity internal to the body that a living being presents itself to others. What is the expression proper to corporeity? What is the language of the body? And, finally, what is the expressivity of animal bodies?

In order to answer these questions, it is necessary to analyse the rich notion of *gesture*, which has a relevant role in Merleau-Ponty's phenomenology, as well as in our reflection about animal expression. In this phenomenology of corporeity, gesture is the primordial and original modality of the être-*au-monde*, a significant movement that opens to intersubjectivity. In other words, a gesture is the visible sign of a corporeal intentionality, of a subjectivity that is, first of all, defined as a bodily potentiality. The gesture of an animal or human body is not just material but it is not yet intellectual, it is rather a meaningful ensemble composed by perception and movement, a meaningful behaviour that is not detachable from its corporeal foundation but does not involve a conscience. According to Merleau-Ponty:

The gestures of behavior, the intentions which it traces in the space around the animal, are not directed to the true world or pure being, but

---

3   Maurice Merleau-Ponty, *Phenomenology of Perception*, trans. by Donald A. Landes (New York: Routledge, 2012), p. 408.

to being-for-the-animal, that is, to a certain milieu characteristic of the species; they do not allow the showing through of a consciousness, that is, a being whose whole essence is to know, but rather a certain manner of treating the world, of 'being-in-the world' or of 'existing.'[4]

The animal body, by its proper presence, has a meaning and opens to a 'corporeal' comprehension. In this perspective, the body is 'a center of actions which radiate over a 'milieu'; it must be a certain silhouette in the physical and in the moral sense; it must be a certain type of behaviour".[5] The presence of a living being radically transforms the physical world outside through its specific and proper *silhouette*, through its corporeal normativity. This kind of existence, this kind of subjectivity always has an orientation: perception is at the heart of the movement, as well as the movement is at the heart of the perception. There is a sort of intimacy that distinguishes a gesture from a sum of physical movements. In other words, there is a normativity that defines the living being:

> Others, which are called living beings, present the particularity of having behavior, which is to say that their actions are not comprehensible as functions of the physical milieu and that, on the contrary, the parts of the world to which they react are delimited for them by an internal norm. By 'norm' here one does not mean a 'should be' which would make it be; it is the simple observation of a preferred attitude, statistically more frequent, which gives a new kind of unity to behavior.[6]

In this way, the 'outside' is authentically inhabited by the organism. The relationship between the organism and the environment is dialectic and expresses the organism's global and qualitative response. In other words, external stimuli are not causes but *occasions* for the organism. Through this mutual correlation between the organism and the environment, the outside becomes a meaningful space, namely a *milieu*. In this context, the notion of milieu or environment is to be conceived in the sense of the *Umwelt*, as famously used by Jakob Von Uexküll. The organism itself composes its worlds. The organism is always oriented, it is a subject equipped with a form of intentionality. To be sure, it is a bodily intentionality, but one that is already meaningful.

---

4   *The Structure of Behaviour*, p. 125.
5   *The Structure of Behaviour*, p. 157.
6   *The Structure of Behaviour*, p. 159.

The openness of the organism has a meaning, a meaning that transforms the geographical environment – detached and objective – into an environment of behaviour (according to the distinction elaborated by Kurt Koffka). In this way, the organism and the environment, defined in isolation, must be substituted with two correlatives: 'the 'milieu' and the 'aptitude,' which are like two poles of behavior and participate in the same structure'.[7]

This analysis brings us to recognize an animal *stylistics*. The bodily gestures of an animal being introduce a *style*, a unique and peculiar way to be in the world. According to Marielle Macé,[8] a style is a characteristic way of tracing a form, a centre of intensity. Each living form creates a peculiar tonality, made of accents, rhythms, pauses: a track of being. What emerge is a veritable *animal semiotics*. Each animal is, in fact, a meaningful openness, a contraction of the dimensions of space and time. As argued by Jean-Cristophe Bailly,[9] the presence of an animal is a *signature*, an ensemble of significant phrases composing a specific and natural *grammar*. The notion of style is also widely adopted by Merleau-Ponty for a significance that is not reducible to a mechanic system of stimulus-response, nor identifiable with an intellectual act. The style, therefore, is a certain manner of organising the dimension of space and time, on which the animal applies its competence. In other words, it is a specific way of radiating its proper presence into the surrounding space. This notion brings us to an original conception of singularity: each animal 'puts a hole' in an undifferentiated space, making it something more: a space-for-the-animal. Each animal is a certain way of being a body, a certain way of living the world: each animal is, in the word of Bailly, the invention of a signature.

Towards an Animal Aesthetics

As we have seen above, the body is the seat of expressivity. The corporeal presence of the living being means *something*, it communicates through a corporeal language. This also applies to animal bodies. The expression of animal bodies passes through the visible dimension:

---

7   *The Structure of Behaviour*, p. 161.
8   Marielle Macé, 'Styles Animaux', *L'Esprit Créateur* 51, 4 (2011), 97-105.
9   Jean-Cristophe Bailly, *The Animal Side,* trans. by Catherine Porter (New York: Fordham University Press, 2011).

animals appear and they produce a *visible form* for others. As everybody knows, Merleau-Ponty's last ontology is entirely focused on the notion of visibility. What is treated here is the *animal visibility*. In the course devoted to animality and animal behaviour, held at Collège de France in 1957-1958, Merleau-Ponty comments on the biological research of Adolf Portmann. What Portmann and Merleau-Ponty share is the valorisation of animal appearance, of the way in which animals show themselves. Against the scientific as much as philosophical prejudice based on the study of the interior, Merleau-Ponty looks for a truth on the surface of the animal forms. The consideration of appearance is not at all uninteresting. According to him (as well as to Portmann):

> There are two ways to consider the animal, as there are two ways to consider an inscription on an old stone: we can wonder how this inscription was traced, but we can also seek to know what it means. Likewise we can either analyse the processes of the animal under a microscope, or see a totality in the animal.[10]

When an animal appears, a spectacle of forms, colours and physiognomies arises. What is fundamental in these analyses is that the animal's appearance is not entirely explicable by the law of economics and utility. The animal form is also made for the eyes looking at it. Portmann compares it to a work of art. According to Portmann and Merleau-Ponty, within the animal world, there is a sort of *useless visibility* that questions the 'Darwinian ideology': 'life is not only an organisation for survival; there is in life a prodigious flourishing of forms, the utility of which is only rarely attested to and that sometimes even constitutes a danger for the animal'.[11] The numerous cases of hypertelia and, in general, the prodigality of forms of nature, demonstrate that appearances do not entirely submit to paradigms of utility and efficacy. Adaptation is not the absolute and unique canon of life. Portmann, and Merleau-Ponty after him, look at the 'spectacle-attitude' of animal bodies: animal bodies are seen, animal bodies are made to be seen. However, denying the ideology of utility and accepting this form of expressivity within the living world does not abolish physical and biological laws: freedom and economy, exception

---

10 Maurice Merleau-Ponty, *Nature. Cours Notes from Collège de France*, trans. by Robert Vallier (Evanston: Northwestern University Press), 2003, p. 187.

11 *Nature,* pp. 185-186.

and law, finalism and determinism: all these elements can co-exist in the living world. According to Merleau-Ponty:

> We must criticize the assimilation of the notion of life to the notion of the pursuit of utility, or an intentional purpose. The form of the animal is not the manifestation of a finality, but rather of an existential value of manifestation, of presentation. What the animal shows is not utility; rather its appearance manifests something that resembles our oneiric life.[12]

The animal form is no longer an accident. Certainly, the relation of causality remains efficacious but it does not exhaust the meaning of this expression. What remains is a sort of power of the body, that is, the power to express something through its visibility. In this way, we can affirm that 'as we can say of every culture that it is both absurd and the cradle of meaning, so too does every structure rest on a gratuitous value, on a useless complication'.[13] There is a sort of contingency that is deeply valorised by Portmann and Merleau-Ponty, a useless complication that opens to the emergence of sense. For this reason, we can define it as a foreign language. Animal appearance produces something for others, it opens to intersubjectivity that is to be understood as intercorporeality. But we have to specify that this visibility is not reducible to actual vision. Animal appearance refers to a *possible eye*, it reveals a presence, it expresses a semantic ensemble. The animal, before being seen, is visible. For this reason, Portmann uses the expression: *self-presentation – Selbst-Darstellung*. There is a priority of visibility on seeing.

In this perspective, the body's appearance is a language: the animal is its own behaviour, its own body. And the animal is also an expression, it owns an expressive value which is completely free and aesthetic. There is always a reference to a possible eye: the body is a semantic whole. But it is not just a bio-semiotic approach, it is also an ontological reasoning. Merleau-Ponty is explicit: the body is an organ of the for-other. This is a central point for his ontology. He affirms, in fact: 'we must grasp the mystery of life in the way that animals show themselves to each other'.[14] The intrinsic visibility of the living being demonstrates that there is an inherent relationality within nature (what Merleau-

---

12   *Nature*, p. 188.
13   *Nature*, p. 188.
14   *Nature*, p. 188.

Ponty – using Husserl's expression – calls *Ineinander*): 'what exists are not separated animals, but an inter-animality'.[15] Defining the body as an *organ of the for-other* underlines a visibility that does not depend on the presence of someone who is actually watching: 'the others' sensoriality is implicated in my own'.[16] This is the reason why each body is already a sort of symbolism. The body sees and is seen, in a general horizon of expression. We have to underline that this expressive and meaningful horizon has an ontological burden: animal appearance is part of Being, of an already-given visibility that is the Being itself. Animal conduct – or way of being a body – is not inevitably addressed to a purpose, it does not submit to an *aut-aut* reasoning. There is virtuality in the natural world. Animal forms are not reduced to utility and adaptation. Animal behaviour and animal appearance are part of an *intra-expression*, of a primordial symbolism: every being sees and is seen. Being and appearing, being and being perceived are the same. This perspective opens an original way for rethinking intersubjectivity: there is a specular relation between animals, 'each is the mirror of the other'.[17] What exists, therefore, is an inter-animality and life is precisely the 'power to invent the visible'.[18] On the one hand, interanimality in the sense of intercorporeity: the body, in fact, is the first openness to alterity, to the others. On the other hand, interanimality in the sense of intervisibility: the body is the mirror of all other bodies, in a horizon of virtual visibility and meaningful expression.

Towards an Animal Symbolism

Adolf Portmann's analysis of animal aesthetics has led us to recognize a contingency inner to life itself. This kind of contingency is not reducible to an accident. On the contrary, it is the reason why we can identify a primordial sense in animal conduct. We have seen that Merleau-Ponty compares this primordial significance to the human oneiric dimension. In order to understand this metaphor, we have to briefly address the work of Konrad Lorenz. Merleau-Ponty comments on it in his course on animality of 1957-1958.

---

15   *Nature,* p. 189.
16   *Nature,* p. 245.
17   *Nature,* p. 189.
18   *Nature,* p. 190.

A student of von Uexküll, Lorenz developed an interesting interpretation of animal behaviour, in particular regarding the notion of *instinct*. In this perspective, Merleau-Ponty finds the way to valorise a primordial symbolism that is inherent to nature. According to Lorenz and Merleau-Ponty, instinct is not a mechanical system of stimulus-response: instinctive tendencies are not actions directed toward a goal. On the contrary, 'instinct is primordial activity 'without object', *objektlos*, which is not primitively the position of an end. [...] The instinct is an activity established from within but that possesses a blindness and does not know its object'.[19] Lorenz provides several examples. In the words of the ethologist:

> I shall never forget the behavior of a starling I kept when I was a schoolboy. One day while sitting on the head of a bronze statue in our dining room, the bird kept scanning the white ceiling in an excited manner. From time to time it took off, flew up to the ceiling, snapped at something, came back to its perch, performed the movements of beating an insect against the perch, then swallowed and appeared satisfied for the moment. I had to climb up on pieces of furniture and even then I had to fetch a ladder before I had really convinced myself that there were no flying insects in that room.[20]

Lorenz defines this kind of conducts as *vacuum activities*. In such cases, the object is put on the back-burner: it becomes evocative of an innate complex, of a *theme*, of a *behaviour pattern*. The external object – external excitant – is the trigger for a sort of 'reminiscence'. According to the lecture of Merleau-Ponty:

> This tension meets the object not so much because it is directed toward it as because it is a means capable of resolving the tension, as if the object intervened like a point of contact that is in the animal, as if it brought to the animal the fragment of a melody that the animal carried within itself, or came to awaken an a priori that provoked a reminiscence.[21]

---

19   *Nature*, pp. 191-192.
20   Konrad Lorenz, *The Foundations of Ethology*, (New York: Springer, 1981), p. 127.
21   *Nature*, pp. 191-192.

Influenced by Raymond Ruyer, Merleau-Ponty speaks about an *experimental Platonism*,[22] leading to the recognising of an *a priori*, that is the *theme* of the animal, what von Uexküll and Merleau-Ponty himself define as 'melody'. These animal behaviours are expressive and sense-making conducts. And they are not the results of an accident, nor a pre-determined plan. The organism is immersed in a horizon of possible and living expressions, of possible significances, of primordial sense that is actualised by the living form. Instinct is directed to a world of image. According to Merleau-Ponty:

> There is an oneiric, sacred, and absolute character of instinct. It seems that the animal both wants and does not want the object. This instinct is both in itself and turned toward the object, it is both an inertia and a hallucinatory, oneiric behaviour, capable of making a world and of picking up any object of the world. [...] Instinct is oriented toward the image or the typical. There is a narcissism of instinct. If it tends to find its identity in fixing an object, it does not know what it is nor what it wants.[23]

Now we can understand why Merleau-Ponty compares the language of animal appearance to oneiric life: as in dreams, there are floating themes in animal behaviour too. The same ambiguity, the same metamorphic dimension: there is a virtuality in the animal world, a veritable pre-culture that does not need any consciousness. The theme is the phantom of the behaviour: the animal behaves following this theme – we can say this *a-priori* – but it is not entirely determined, because the *floating theme* leaves room for a *natural freedom*: the theme does not end with its actualisation. In this perspective, the object of the instinctive activity is not its cause, but a sort of foothold of the animal's theme, an abstract trigger-stimulus. According to Lorenz and Merleau-Ponty, therefore, instinctive conduct has dramatic aspects, 'a vital drama from the moment that being is both vision and passion, when being carries both the internal law of its behavior and the relations to an exterior world'.[24] In this way, Merleau-Ponty connects these empty activities to a primordial form of symbolism:

---

22 Maurice Merleau-Ponty, *Institution and Passivity: Course Notes from the Collège de France 1954-1955*, trans. by Leonard Lawlor and Heath Massey (Evanston: Northwestern University Press, 2010), p. 17.
23 *Nature*, p. 193.
24 *Nature*, p. 194.

With empty activity, instinct is going to be capable of being derailed or is going to pass from instinctive activity to symbolic activity. Empty or outlined activities are going to become means of communications for the animals.[25]

Animal conduct is directed toward the *typical*, toward an imaginary horizon. Animals live in an expressive, significant and symbolic horizon: corporeity, visibility, aesthetics and symbolism are intertwined. Merleau-Ponty adopts psychoanalytic vocabulary in order to define the animal's 'vital drama', in order to define an expressive horizon in which presence and absence, internal and external world, vision and passion can coexist. The ritualization and natural duplicity of the instinct lead us to recognize a primordial narcissism inherent in animal behaviour.

The development of instinct into symbolic functions is inscribed in the way in which instinct is constituted because it is *objektlos* [without object], and from this fact, it possesses an imaging function. Behaviours instituted by the development of symbolism acquire a new value as social evocation. [...] There is a 'miming exaggeration'. We used the term 'ritualization' concerning this.[26]

The ritualization is possible because the object of the behaviour is first of all an image, because there is a distance between doing and seeing. This is a fundamental point in our reasoning: in this distance there is room for the emergence of a meaningful expression and communication, of a corporeal 'dialogue'.[27] According to Merleau-Ponty:

We can also bring these facts of the phenomena close to human language. As Lorenz says: 'just as different forms of verbal symbols of our language are not ordered by their signification and are fixed only by convention, so too is it for the innate social trigger schemes'. We see a ceremony emerge across facts in a very narrow and even mechanical dependence on instinct, a thrust of symbolism that uses the facts. Just as the signification of our verbal concepts can be developed into different significations, sometimes even opposed to each other, [...] so too does behavior take on different significations: in one species of Cyclades fish, the behavior that originally indicates inferiority assumes the

---

25   *Nature*, p. 195.
26   *Nature*, p. 197.
27   *Nature*, p. 198.

meaning of a threatening behavior for the dwarf Cyclades. This is why Lorenz proposes the study of a 'compared philology' of the triggers of behavior.[28]

Tracing narcissistic modalities in animal behaviour leads us to an interesting comprehension of animality. In animal behaviour – understood as significant and irreducible; in the animal body – understood as situated and expressive; and finally in animal instinctual activity – understood as orientation and ritualization, there is an 'empty production', a primordial symbolism and pre-culture that are immanent to the body itself. And there is a primordial communication between the body and the world, a communication in the terms of a *tacit language*. Equally, different bodies can communicate and understand each other through this primordial symbolism. It is here – in this *inter-corporeality*, in this *inter-world* – that we find the 'strange kinship between the human and the animal'.[29] It is the investigation into corporeity that leads us to authentically comprehend animal and human beings: no longer separated by an ontological abyss, no longer assimilated by anthropomorphic characteristics. A new kind of *difference* appears here: the human being is not an *exception* but he emerges from an ontological continuity; otherwise, this kind of continuity is not explicable with a biological continuum, as if the human being was at the top of a straight and hierarchic path. The investigation into corporeity leads us to the primordial relationship between animals and humans: first of all, we are a living body, situated and open to the world; but on the other hand, there are differences. The animal, in its *being-body*, is an individual existence equipped with a uniqueness that is expressed in its behaviour, in its activity. The ontology traced by Merleau-Ponty is truly relational: human and animal beings, through their specific bodies, live the same world and share a primordial openness to it. The relation between animals and humans is a *lateral* one.[30] From this perspective, the human being is no longer the result of a mere addition, but *another way of being a body*:

> The relation of the animal to the human will not be a simple hierarchy founded on an addition: there will already be another manner of being a body in human being. We study the human through its body in order

---

28   *Nature*, p. 198.
29   *Nature*, p. 214.
30   *Nature*, pp. 211, 268, 273.

to see it emerge as different from the animal, not by the addition of reason, but rather, in short, in the *Ineinander* with the animal (strange anticipations or caricatures of the human in the animal), by escape and not by superposition.[31]

Conclusion. Towards an Incorporated Meaning

The Merleau-Pontyan ontological perspective I have drafted in this article leads to the recognition of a primordial sense that is already present and expressed in animal appearance and animal conduct. What emerges is a new way to conceive intersubjectivity, as an expressive communication that passes through intercorporeity and intervisibility. The meaning characterising animal forms represents a new way to conceive significance: neither materiality, nor thought. In Merleau-Ponty's perspective, meaning is always incorporated: what we are looking for is the possibility to find other incorporations of meaning not exclusively human. In other words, what we are looking for is an intertwining between *physis* and *logos*, a continuity between the 'silence' of nature and the spoken language of human beings. And this is exactly the research conducted by Merleau-Ponty in the last period of his philosophy, characterised by the so-called *ontological turn*.

Merleau-Ponty, in fact, defines animality as 'the logos of the sensible world: an incorporated meaning'.[32] This definition is fundamental to us: the symbolism of human language is not detached from natural symbolism, from this primordial emergence of significance we have underlined in animal activities. According to him:

> The invisible, mind, is not another positivity: it is the inverse, or the other side of the visible. We must retrieve this brute and savage mind beneath all the cultural material that is given. Here the title takes on its whole meaning: *Nature and Logos*. There is a Logos of the natural aesthetic world, on which the Logos of language relies.[33]

That does not mean that there are not differences between the dimension of the natural symbolism and the linguistic one. Here, we

---

31   *Nature*, p. 214.
32   *Nature*, p. 166.
33   *Nature*, p. 212.

have to briefly deepen the ontology proposed by Merleau-Ponty. In this kind of dimensional ontology, each new layer is an emergence, an introduction of a new architectonic that does not abolish the previous one. This applies to humanity, that is, as already mentioned, a new and different corporeity.

In this way, we can admit a primordial symbolism and a natural and bodily communication, a tacit language that is, however, also *expressed* by animal bodies. The expression of the natural world, in which we have found animal visibility, participates of the articulation of meaning. Meaning is not exclusive to the human being, because it belongs to the Being. We can recognize the meaning of the body because the Being itself is productive of meaning.

Finally, it is in this intertwining between natural expression and human language that we can reformulate an original conception of meaning: a sense always incorporated and appearing; it is in this intertwining that we can reformulate the notion of human enunciation, never detached from natural world and from animal expression; it is in this intertwining that we can rethink the intersubjectivity which is able to include animal bodies in a general horizon of expressivity, visibility and bodily communication. And this is precisely the ontological horizon Merleau-Ponty was formulating in *The Visible and the Invisible*:

> No one thinks any more, everyone speaks, all live and gesticulate within Being, as I stir within my landscape, guided by gradients of differences to be observed or to be reduced if I wish to remain here or to go yonder. Whether in discussion or in monologue, the essence in the living and active state is always a certain vanishing point indicated by the arrangement of the words, their 'other side', inaccessible, save for him who accepts to live first and always in them. As the nervure bears the leaf from within, from the depths of its flesh, the ideas are the texture of experience, its style, first mute, then uttered. Like every style, they are elaborated within the thickness of being and, not only in fact but also by right, could not be detached from it, to be spread out on display under the gaze.[34]

---

34   Maurice Merleau-Ponty, *The Visible and The Invisible*, trans. by Alphonso Lingis, (Evanston: Northwestern University Press, 1968), p. 119.

Bibliography

Bailly, Jean-Cristophe, *Le versant animal* (Paris: Bayard, 2007), trans. by Catherine Porter, *The Animal Side* (New York: Fordham University Press, 2011).

Barbaras, Renaud, *De l'être du phénomène. Sur l'ontologie de Merleau-Ponty* (Grenoble: Jérôme Millon, 1991), trans. by Ted Toadvine and Leonard Lawlor, *The Being of the Phenomenon. Merleau-Ponty's Ontology* (Bloomington: Indiana University Press, 2004).

Burgat, Florence, *Liberté et Inquiétude de la vie animale* (Paris: Editions Kimé, 2006).

Burgat, Florence, *Une autre existence. La condition animale* (Paris: Albin Michel, 2012).

Fóti, Véronique M., *Tracing Expression in Merleau-Ponty. Aesthetics, Philosophy of Biology, and Ontology* (Evanston: Northwestern University Press, 2013).

Goldstein, Kurt, *Der Aufbau des Organismus. Einführung in die Biologie unter besonderer Berücksichtigung der Erfahrungen am kranken Menschen* (Den Haag: Nijhoff, 1934), trans. by Kurt Goldstein, *The Organism. A Holistic Approach to Biology Derived from Pathological Data in Man* (New York: Zone Books, 1995).

Hardouin, Robert, *Le mimétisme animal* (Paris: Presses Universitaires de France, 1946).

Koffka, Kurt, *The Principles of Gestalt Psychology* (London: Routledge, 1935).

Lorenz, Konrad, *The Foundations of Ethology* (New York: Springer, 1981).

Macé, Marielle, 'Styles Animaux', *L'Esprit Créateur* 51, 4 (2011), 97-105.

Merleau-Ponty, Maurice, *La Structure du comportement* (Paris: Presses Universitaires de France, 1942), trans. by Alden L. Fisher, *The Structure of Behaviour* (Boston: Beacon Press, 1967).

Merleau-Ponty, Maurice, *Phénoménologie de la perception* (Paris: Gallimard, 1945), trans. by Donald A. Landes, *Phenomenology of Perception* (New York: Routledge, 2012).

Merleau-Ponty, Maurice, *L'Institution, la passivité: notes de cours au Collège de France* (Paris: Editions Belin, 2003), trans. by Leonard Lawlor and Heath Massey, *Institution and passivity: course notes from the Collège de France 1954-1955* (Evanston: Northwestern University Press, 2010).

Merleau-Ponty, Maurice, *Le visible et l'invisible* (Paris: Gallimard, 1964), trans. by Alphonso Lingis, *The Visible and The Invisible* (Evanston: Northwestern University Press, 1968).

Merleau-Ponty, Maurice, *La Nature. Notes. Cours de Collège de France (1956-1960)* (Paris: Seuil, 1995), trans. by Robert Vallier, *Nature. Cours Notes from Collège de France* (Evanston: Northwestern University Press, 2003).

Portmann, Adolf, *Die Tiergestalt, Studien über die Bedeutung der tierischen Erscheinung*, 2nd ed., (Basel: Reinhardt, 1960), trans. by Hella Czech, *Animal Forms and Patterns: A Study of the Appearances of Animals* (New York: Schocken Books, 1967).

Toadvine, Ted, 'How not to be a Jellyfish. Human Exceptionalism and the Ontology of Reflection', *Phenomenology and the Non-Human Animal*, ed. Corinne Painter and Christian Lotz (Doordrecht: Springer, 2007), 39-55.

Toadvine, Ted, *Merleau-Ponty's Philosophy of Nature* (Evanston: Northwestern University press, 2009).

Uexküll, Jakob von, *Umwelt und Innenwelt der Tiere* (Berlin: Springer, 1909).

Uexküll, Jakob von, *Streifzüge durch die Umwelten von Tieren und Menschen – Ein Bilderbuch unsichtbarer Welten* (Berlin: Springer, 1932).

Westling, Louise, *The Logos of the Living World. Merleau-Ponty, Animals, and Language* (New York: Fordham University Press, 2014).

Richard Hutchins

MAKING THE ABSENT REFERENT PRESENT: THE SACRIFICE OF WOMEN AND ANIMALS IN LUCRETIUS

In one of the earliest works of scholarship to provide a feminist framework for linking the abuse of animals with the abuse of women ('Woman-Battering and Harm to Animals', 1995), Carol Adams claims that in patriarchy it is not so much that men have power over women and humans in general have power over animals.[1] Rather, men have power over women, (feminized) men, and (feminized) animals. In other words, male power pervades nature in gendered form. Adams was one of the first to notice that descriptions of animal abuse are often laced with connotations of gender: abused animals tend to be gendered as female, and the act of animal abuse is described in ways similar to the abuse of women.

'Gender', Adams says, 'is an unequal distribution of power; interconnected forms of violence result from and continue this inequality. In a patriarchy', Adams continues, 'animal victims, too, have become feminized. A hierarchy in which men have power over women and humans have power over animals, is actually more appropriately understood as a hierarchy in which men have power over women, (feminized) men, and (feminized) animals'.[2]

This chapter locates a precursor to Adam's ecofeminist framework, the 'absent referent', in the Epicurean poet and philosopher Lucretius

---

1   I am grateful for the hospitality and intellectual generosity of the organizers of *Bestiarium: Rappresentazioni dell'umano e dell'animale* at the University of Verona, and for the work of the two anonymous reviewers. I am also grateful to Yung In Chae for numerous discussions about origins in Foucault, Agamben, and Said, as well as to Alex Petkas for going out of his way, when he himself had no time, to give me his always sharp suggestions about Greek and Latin. Such friends make it a joy to write.

2   Carol Adams, 'Woman-Battering and Harm to Animals', in *Animals and Women: Feminist Theoretical Explorations*, ed. by Carol Adams and Josephine Donovan (Durham: Duke University Press, 1995), pp. 55-84 (p. 80).

and his epic poem, *De rerum natura*, or *On the Nature of Things*.[3] The absent referent is the idea that descriptions of violence against women often contain references to animals, and that descriptions of violence against animals are often gendered so as to contain references to women. While Adams is primarily interested in the ways that meat eaters conceal the animality behind the meat they eat, she also shows how meat eating has come to be associated with masculinity and the butchery and objectification of female bodies. In short, Adams thinks, violence against women tends to be described in terms of violence against animals and violence against animals in terms of violence against women.

This chapter will show how Lucretius' description of male violence against women and animals resonates with Adams' notion of the absent referent through political close readings of two interconnected texts from *De rerum natura*. The first text is from Book One. It describes the sacrifice of Iphigenia, which is depicted as an animal sacrifice. The second text is from Book Two of *De rerum natura*, and shows a calf being sacrificed on an altar, while its grief-stricken mother wanders in search of it. These are two of the most moving scenes in *De rerum natura*. In the latter passage about the sacrificed calf, I argue that Lucretius makes the absent referent present by turning the sacrifice of Iphigenia on its head. That is, while the sacrifice of Iphigenia is described in terms that connote animal sacrifice, the sacrifice of the calf is described in terms that connote human sacrifice. Hence, by reading the two passages together, I show that Lucretius makes the absent animal referent in the sacrifice of Iphigenia present again in the sacrifice of the calf, while the human referent in the sacrifice of Iphigenia is forced to play the role of the absent referent in the sacrifice of the calf.

The methodology of this chapter, therefore, consists of putting these two scenes from *De rerum natura* into a hermeneutic circle, to point out their resonances and to suggest reasons for thinking that Lucretius meant the two passages to be read in dialogue with each other. An important claim in this chapter, therefore, is that the sacrifice of Iphigenia in Book One foreshadows not generally or vaguely but quite specifically the sacrifice of the calf in Book Two; and that the sacrifice of the calf in Book Two is foreshadowed in such a way that it cannot be read without thinking about the traces of Iphigenia from Book One. Once

---

3   Lucretius' dates are roughly 99-55 BCE. He was a contemporary of Cicero,
     and lived in the last century of the Roman Republic. *De rerum natura*, or *On
     the Nature of Things*, was left unfinished at his death.

this hermeneutic circle is set in place, these passages reveal Lucretius to have a strikingly contemporary perspective on the intersectional nature of patriarchal violence against women and animals. Since it is important for a critical movements like the environmental humanities to identify its historical precursors, one task of this chapter is to present Lucretius as a intellectual forerunner – with all the complications and misrecognitions this brings – for ecofeminist critiques of patriarchal culture. Locating such intellectual ancestors is doubly important, since ecofeminism is just now emerging for a second time as a mode of critique relevant to an era rife with ecocide and misogyny.

Unlike Adams' more analytical form of criticism in *The Sexual Politics of Meat*, which was itself a development out of deconstruction,[4] the absent referent in Lucretius arises more from Lucretius' tendency to use myth and mythmaking as ways to critique Roman culture. The sacrifice of Iphigenia and of the calf are, in fact, origin or foundation myths. They are not just myths of violence but myths of foundational or 'originary violence'. According to Cary Wolfe, originary violence is the way that the law (and the violence sedimented in the law) 'installs its frame for who's in and who's out' of the human community, and for who gets its protections. In Wolfe's view, the law installs its frame by establishing origins that justify what is killable and what is not.[5] But where, for Wolfe, originary violence usually takes the form of myths that *justify* violence, Lucretius' myths *critique* violence against women and animals under male-dominated *religio*. While Adams calls this structure patriarchy, Lucretius has no such word. Instead, he calls it *religio* ('religion'/'superstition'). Since religion is

---

4   Cf. Carol Adams, *The Sexual Politics of Meat: A Feminist-Vegetarian Critical Theory* (London: Bloomsbury, 2010 [1990]), pp. 5-6.

5   Cf. Cary Wolfe, *Before the Law: Humans and Other Animals in a Biopolitical Frame* (Chicago: University of Chicago Press, 2013), pp. 8-9. Wolfe elaborates on his conception of originary violence this way: 'And here – to move to the main part of my title – we can begin to glimpse the many senses of what it means to be 'before the law': 'before' in the sense of that which is ontologically and/or logically antecedent to the law, which exists prior to the moment when the law, in all its contingency and immanence, enacts its originary violence, installs its frame for who's in and who's out. This is the sense of 'before' that is marked by Arendt's speculations on the 'right to have rights,' and it is against such a 'before' that the immanence of the law and its exclusion is judged. And thus, 'before' in another sense as well, in the sense of standing before the judgment of a law that is inscrutable not just because it establishes by fiat who falls inside and outside the frame, but also because it disavows its own contingency through violence'.

not often a focus of Adams' work, perhaps Lucretius' ancient perspective has something to offer ecofeminism in terms of locating the origins of structural violence against women and animals not in patriarchy generally but specifically in patriarchical religion. Lucretius, as we will see, could not be clearer that the origin of this violence resides not with men alone but with men using religion for the sake of power.

It is important to note that *religio* in Lucretius does not mean 'religion' exactly. Nor is it entirely different either. For Lucretius, *religio* refers to any addition of thought or opinion to the bare system nature. It is an imposition of human thought and meaning onto an indifferent, nonhuman world. Whether we call this imposition 'myth', 'ideology', or 'superstition' (the latter is often the preferred translation for *religio* in Lucretius),[6] the problem with *religio*, Lucretius thinks, is that it offers tools for the powerful (almost always men) to do violence to the rest of the living world. Lucretius' myths, therefore, are critical myths rather than justificatory ones.

Reading closely the sacrifice of Iphigenia and of the calf in dialogue with one another will reveal that Lucretius is not just critiquing religion in general, but highlighting the specific kinds of victims who suffer systematically under male-dominated *religio*. Lucretius' representation of women and animals, therefore, comes very close to what another philosopher-critic famously called 'the non-power at the heart of power'.[7]

---

6   'Superstition' is the preferred translation in the commentary by Leonard and Smith: William Elley Leonard and Stanely Barney Smith, *T. Lucreti Cari De rerum natura libri sex* (Madison: University of Wisconsin Press, [1942] 1968), p. 206. And yet, the character Balbus in Cicero's *De natura deorum* (*On the Nature of the Gods*) distinguishes between *religio*, which for him is a positive term that denotes the understanding of the true meaning of myth and ritual, and *superstitio*, which Balbus considers a negative term, denoting popular or literal and unquestioning belief in received myths. Cf. Monica Gale, *Myth and Poetry in Lucretius* (Cambridge: Cambridge University Press, [1994] 2007), p. 25, note 90. What Lucretius criticizes about *religio*, therefore, is not just uncritically believing certain things about the gods but also the *consequences* of such belief, which he characterizes as something that 'weighs down' (*gravi sub religioni, De rerum natura* 1.63), 'oppresses' (*oppressa*, 1.63), and 'stands on top of' or 'crushes' (*super…instans*, 1.65) the human spirit. Epicureans, for their part, advocated a notion of the gods that hewed as closely as possible to the bare realities of nature, as they thought them to be, i.e. atomic nature. Epicureans were not, as is sometimes popularly thought, atheists. They believed in gods, but in naturalized, material, atomic gods, who dwell in tranquility in the *intermundia*, i.e. in the empty spaces between worlds.

7   Jacques Derrida, *The Animal That Therefore I Am*, trans. by David Wills (New York: Fordham University Press, 2008).

The Absent Referent: Animal Traces in the Sacrifice of Iphigenia

In the most famous scene from Lucretius' *De rerum natura*, Agamemnon, the commander of the Greek navy, sacrifices his own daughter, Iphigenia – or 'Iphianassa', as Lucretius calls her – to receive favorable winds from the goddess Diana to sail to Troy. As classicists have noted, the sacrifice of Iphigenia is described as both an inverted wedding and as an animal sacrifice.[8] Connotations of animality, i.e. the way the animal becomes the absent referent in Iphigenia's sacrifice – and not connotations of marriage, although important – will be the focus here. In the passage below, Lucretius famously uses the sacrifice of Iphigenia to warn us about the dangers of religion.

Illud in his rebus vereor, ne forte rearis	80
impia te rationis inire elementa viamque	81
indugredi sceleris. quod contra saepius illa	82
religio peperit scelerosa atque impia facta:	83
Aulide quo pacto Triviai Virginis aram	84
Iphianassai turparunt sanguine foede	85
ductores Danaum delecti, prima virorum.	86
cui simul infula virgineos circumdata comptus	87
ex utraque pari malarum parte profussast,	88
et maestum simul ante aras adstare parentem	89
sensit et hunc propter ferrum celare ministros	90
aspectuque suo lacrimas effundere civis,	91
muta metu terram genibus summissa petebat.	92
nec miserae prodesse in tali tempore quibat	93
quod patrio princeps donarat nomine regem;	94
nam sublata virum manibus tremibundaque ad aras	95
deductast, non ut sollemni more sacrorum	96
perfecto posset claro comitari Hymenaeo,	97
sed casta inceste nubendi tempore in ipso	98
hostia concideret mactatu maesta parentis –	99
exitus ut classi felix faustusque daretur.	100
tantum religio potuit suadere malorum.	101

8 Cf. Cyril Bailey, *T. Lucreti Cari De rerum natura*, 3 vols. (Oxford: Oxford University Press, [1947] 2001), vol. 2, pp. 614-15; and Hugh Andrew Johnstone Munro, *Titi Lucreti Cari De Rerum Natura Libri Sex, with a Translation and Notes* (Cambridge: Cambridge University Press, 1846) ad loc., pp. 126-8.

The one thing I fear in this matter is that you perhaps think that you are entering into the principles of an unholy system (*impia rationis elementa*) and a path of crime. But, on the contrary, too often religion has given birth to wicked and unholy deeds: as when at Aulis the chosen leaders of the Danaans, the prime of men, foully defiled the altar of the Virgin of the Crossroads with the blood of Iphianassa (i.e. Iphigenia). As soon as the ribbon had bound her virgin braids flowing down equally on both sides of her cheeks, and as soon as she sensed her sad parent standing at the altar and next to him the attendants hiding the knife and her countrymen pouring out tears at the sight of her, mute with dread (*muta metu*), lowering herself (*summissa*) by her knees she reached for the ground. It was no help to the poor girl that she was the first to have given (*donarat*) the name of 'father' to the king; for lifted up by the hands of men, she was led to the altar trembling, not so that when the solemn custom of the ritual was completed she might be escorted by the loud wedding song, but so that as a pure victim (*casta hostia*), sad, she might fall impurely (*inceste*) by the sacrificing stroke of her father at the very age of marriage – so that a happy and fortunate departure might be given to the navy. So powerful was religion in persuading (literally 'making sweet') evils.

Here, Lucretius argues that serious dangers arise from having false views about nature: in the case of Iphigenia, from believing that gods intervene in the world and that, therefore, humans should give them worship, and even human sacrifice.[9] But since Lucretius believes, contrary to Roman religious norms, that gods do not intervene or act in the world, sacrifice, in Lucretius' view, can only be a pointless cruelty, whether of humans or of animals. Lucretius describes the sacrifice of Iphigenia as both an inverted wedding and as an inverted animal sacrifice, and numerous ironies follow from this. First, whereas Lucretius thinks that a true knowledge of nature would be life producing, religion, for him, has a deadly fertility, as he says: 'religion has *given birth* to wicked and impious deeds' (*religio* peperit *scelerosa atque impia facta*, line 83). Religion, therefore, for Lucretius, turns out to be a kind of necropolitics, opposed to the vitalism of animal life and of human marriage, both of which are celebrated throughout *De rerum natura*.

As classicist Monica Gale has noted, Lucretius specifies the agents of this religious violence as males, and characterizes them, ironically, as the 'chosen leaders of the Danaans, the prime of men'

---

9   David Sedley, *Lucretius and the Transformation of Greek Wisdom* (Cambridge: Cambridge University Press, [1998] 2003), p. 154, note 37.

(*ductores Danaum* delecti, prima *virorum*, line 86).[10] They are, indeed, 'chosen' and 'primary' leaders – but of a pointless cruelty. Instead of producing life, they 'foully defiled with Iphigenia's blood the altar of Diana' (*Triviai Virginis aram/Iphianassai turparunt sanguine foede*, line 85). Diana was originally a tree spirit, the goddess of wilderness, purity, the moon, and wild animals.[11] Lucretius also specifies that it was 'the hands of men' (virum *manibus*, line 95) by which Iphigenia was 'led' (*deductast*, line 96) to the altar against her will. Such hands, ironically, signified trust or protection in paternalistic Roman culture. The leadership of these men is, therefore, no leadership at all. They lead Iphigenia like a scared and unwilling animal to slaughter, whose affect – for she has no voice – can only be articulated through her trembling body (*tremibunda*, line 95).[12]

Iphigenia's affect also blends with the narrator's voice: that is, in the striking repetition of 'ah' sounds in *tremibundaque ad aras/deductast*, which operates as a kind of emotional release-point. These sounds seem to signal a freely-circulating feeling or emotion, that is shared not only with the internal audience of weeping citizens – the *civis* of line 90, who 'pour out their tears' (*lacrimas effundere*) – and not only with the energies and rhythms of 'Lucretius" poetry, but also with the free-floating, affective atmosphere in the passage in general.[13] The 'ah' sounds seem to mark the emotions that accompany the revelation that the narrative logic of the passage has been leading up to – when the truth of *religio* has finally been revealed: 'ah ah ah ah!'[14]

---

10 Gale [1994] 2007.

11 Leonard & Smith, ad loc, p. 212.

12 Munro notes that *tremibunda* ('trembling') expresses at once the trembling of the animal victim and the fluttering anxiety of a bride. Cf. Munro, p. 127.

13 Cp. what Julia Kristeva says about bodily affect as the basis of symbolic affect in literature, in *New Maladies of the Soul*, trans. by Ross Guberman (New York: Columbia University Press, 1995), p. 104: '[There is] the *semiotic*, which consists of drive-related and affective *meaning* organized according to primary processes whose sensory aspects are often nonverbal (sound and melody, rhythm, color, odors, and so forth), on the one hand, and *linguistic signification* that is manifested in linguistic signs and their logico-syntactic organization, on the other'.

14 For the idea that different narrative structures or logical flows, whether in literary or philosophical texts, can release new or different affects, see Deleuze's reading of affect in Spinoza. Cf. Gilles Deleuze, 'Spinoza and the Three 'Ethics", in *Essays Critical and Clinical*, trans. by Daniel W. Smith and Michael A. Greco (Minneapolis: University of Minnesota Press, 1997), pp. 138-51.

Connected with the theme of affective release-points and revelation is the theme of 'binding' that links religious violence against women and against animals. For instance, Lucretius describes Iphigenia as having a ribbon, or *infula*, 'bound' (*circumdata*, line 87) around her 'virgin *braids*' (*virgineos…comptus*, line 87). This *infula*, or ribbon, blindfolds her. The binding also looks to the folk etymology of *religio*, which was thought to derive from *re-ligare*, 'to bind down' or 'to bind back.' Iphigenia is, in fact, bound with a sacrificial animal ribbon, an *infula*, rather than a marriage ribbon, a *vitta*.[15] It is this animal binding that ties Iphigenia's 'virginally bound (*virgineos…comptus*) hair', which has been prepared as if for marriage. These human-animal bindings connect to the larger theme of binding between humans and gods that Iphigenia's sacrifice sets in motion. There is, for instance, the ambiguity of *quo pacto* at the beginning of the passage (line 84): *quo pacto* can mean both 'in which way' and 'by which *pact*'. Iphigenia is bound to death under this pact (*quo pacto*), so that the gods will give the navy winds to sail to Troy (*exitus ut classi felix faustusque* daretur, line 100). It is through the motif of binding that violence against women and animals is, in Lucretius' view, structurally bound up with male-dominated *re-ligio*.[16]

Through the binding power of *religio* – sightless, and like an animal led to slaughter – Iphigenia is reduced to sensation alone (*sensit*, line 90). She goes from human to scapegoat, suspended in a state of humananimality. This suspension is marked by Lucretius as a loss of humanity, and contributes to passage's sense of a tragic double-bind. *Religio*, in fact, turns Iphigenia into 'bare life'.[17] She only *senses* the

---

15   Cf. Leonard & Smith, p. 210.
16   For the practice of adorning the sacrificial animal with a ribbon, or *infula*, and other binding practices, Cf. Jan Bremmer, 'Greek Normative Animal Sacrifice', in *A Companion to Greek Religion*, ed. by Daniel Ogden (Wiley-Blackwell, [2007] 2010), 132-144 (pp. 134-5); and John Scheid, 'Sacrifices for the Gods and Ancestors', in *A Companion to Roman Religion*, ed.   by Jorg Rupke (Wiley-Blackwell, [2007] 2011), 263-71 (p. 264).
17   For 'bare life' see Giorgio Agamben, *Homo Sacer: Sovereign Power and Bare Life*, trans. by Daniel Heller-Roazen (Stanford University Press, 1998) p. 9. Bare life is neither *bios*, which is life in a collectivity or political community; nor is it *zoe*, which is unqualified living existence, as *zoon* denotes in Greek. *Zoe* exists prior to language and community. *Bios* exists only in community. Bare life, however, is a third term. It is what comes into being when *zoe* enters into *bios*, i.e. when unqualified life (*zoe*) is on the threshold of entering into or forming community, and begins to take on qualities associated with living in a community, like politics and language. More dramatically, however, bare

sadness of her murderous father, Agamemnon, standing by the altar (*maestum...parentem/sensit*, line 89-90). And yet, her feelings seem to remain value-laden and to contain content. For example, she is able to sense the fact that, as Lucretius says, the priests are concealing the sacrificial knife nearby her.[18] The word for 'knife' here is *ferrum* (literally 'iron', line 90), which tends to be associated by Lucretius with *ferus*, 'wild'/'barbaric'/'feral'. Given the atmosphere of the passage, Lucretius seems to provoke these associations in his readers. Iphigenia also senses her fellow countrymen's gaze (*civis*, line 91). And yet, there is an irony in the fact that these countrymen (or 'citizens', an alternate translation for *civis*) pour out tears 'at the sight of her' (*aspectu suo*, line 91). Like Lucretius' readers, these *civis* are privileged to see the horror that Iphigenia, the victim, cannot. *aspectu suo*, therefore, points to a layered dramatic irony – a dramatic irony not only in respect to us, the readers, but also in respect to the internal audience of *civis* watching the sacrifice: sympathetic and perhaps partisan to her. And yet, Iphigenia, because she cannot be sure of what is going on around her on account of sensory deprivation, lacks even the basis for the sadness that Lucretius says her father and her fellow *civis* feel. While it is true that Iphigenia is called 'sad' (*maesta*, line 99), this is an appellation that belongs less to Iphigenia's inner, emotional world than to what Bakhtin calls the voice of 'going opinion,' i.e. the voice of society as transmitted through myth. It is therefore the voice of myth/society ventriloquized through the voice of the narrator that calls Iphigenia 'sad'.[19] This is because it is

---

life also comes into being when those already living in communities (*bios*) are violently removed from that community, and lose the protections, rights, status, and language of that community. It is important to note that when cast out, *bios* does not become *zoe* again, in Agamben's view. It does not return to being unqualified life. And yet it is no longer *bios* either, since it has been cast out of the community. Bare life is, therefore, neither completely unqualified life, nor political life, but something in the interzone between the two. Bare life, therefore, tends to be difficult to classify. The important fact is that by becoming bare life *bios* becomes killable with impunity. And more horrific still, it is precisely in rejecting bare life that a community founds itself a community in the first place.

18   For the practice of concealing the knife from the animal victim, cf. Bremmer, p. 136.

19   For the voice of the 'common view', 'a given sphere of society', 'going opinion', or 'current opinion', cf. Mikhail Bakhtin's famous reading of Dickens' *Little Dorrit* in, 'Discourse in the Novel', in *The Dialogic Imagination: Four Essays* (Austin: Texas University Press, 1981), pp. 259-422 (pp. 301 and 305).

Iphigenia's emotional ill-luck to be stripped of even the possibility of feeling the emotions proper to a human being in such situation. While objectively 'sad', because hers is a sad myth, in Lucretius' telling, she is less sad than confused or in suspension.

Like a dumb animal 'mute with dread' (*muta metu*, line 92), Iphigenia can only rely on her value-laden sensations to register the impending violence. Her voice has been removed, as emphasized by the 'm' sounds in *muta metu*, which may imitate non-articulate animal sounds. The interchangeability of the letters and the chiastic structure of the vowels in *mUtA mEtU* also suggest an interconnection between her fear and her silence. As the scene comes to its conclusion, Lucretius says that Iphigenia, 'lowered herself by her knees' (*summissa genibus*, line 92) and reached out for (or 'petitioned') the earth (*terram petebat*, line 92). If *petebat* contains a sense of 'petition', then the phrase *terram petebat* might suggest a plea to the earth (*terra mater*) against the violence of the sky gods, mentioned just prior to this passage at *De rerum natura* 1.63-5.[20] Iphigenia's disorientation and unwillingness to proceed to the altar would, in the normal practice of Greek or Roman sacrifice, have nullified the sacrifice, even were she an animal. This is because according to Greco-Roman custom, animals were supposed to have gone willingly to the altar. Iphigenia does not. Animal victims were also supposed to have lowered their heads to indicate consent.[21] Iphigenia simply falls to the ground.[22]

In addition to specifying that it was male hands that did the deed (line 95), Lucretius also underlines the masculine nature of religious violence

---

20 Consider the famous pun leading up to this passage (1.62-5), in which Lucretius etymologizes the origin of *religio* by linking it to faces that appear in the 'region of the sky' (*caeli regione*), and which 'press down upon' (*oppressa*) human life on earth:
    Humana ante oculos foede cum vita iaceret/in terris oppressa gravi sub RELIGIONE,/quae caput a caeLI REGIONE ostendebat (cp. *RE-LI-GIONE* vs. *caeLI-REGIONE*)/*horribile super aspectu mortalibus instans....*
    'When human life lay foully oppressed on the earth, crushed under RELIGION, which displayed a head from the REGION of the heavens, standing over mortals with a horrible face.' It is against this oppression of the sky gods that Iphigenia can perhaps be seen as 'seeking' or 'petitioning' mother earth below (*terram petebat*) as she falls to her knees.

21 Cf. Bremmer, p. 135; Burkert, p. 56; and Scheid 2003, p. 83.

22 Munro suspects that *summissa genibus* ('let down by knees') may refer to the knees of others, i.e. to the knees of the male priests: they hit her with their knees to make her fall. Cf. Munro, p. 127.

by saying that, 'she (Iphigenia) was the first to have given the name of 'father' (*patrio nomine*) to the 'king' (*regem*, line 94)'. In saying that Iphigenia was the first to have 'given' (*donarat*, which means 'given as a gift') the name of 'father' to Agamemnon, Lucretius strongly hints that Agamemnon is violating the familial bond of reciprocity with his own daughter. Instead of reciprocating with her by giving her a gift in return for her having given him the 'name of king', Agamemnon breaks out of the familial reciprocal bond and gives Iphigenia as a gift to Diana to establish reciprocity with the gods. This bond with the gods is entered into for deadly, anti-nature purposes: so that winds might be given to the navy to sail to Troy (*daretur*, line 100). Not only is this a clear violation of familial reciprocity between father and daughter – one substituted by a pointless and cruel bond with the gods – but it is also an entry into a religious reciprocity whose aim is war. The navy will sail to Troy to begin a decade's long campaign.[23] In a final outrage at the hands of male-dominated *religio*, however, Lucretius calls Iphigenia a *casta hostia* (lines 98-99), a 'pure' or 'unblemished sacrificial victim', who, as he specifies, is killed by the stroke of her own father, *mactatu parentis*. It was not required in ancient sacrifice that the king himself strike the blow, but Agamemnon did.

Not only has Lucretius presented Iphigenia's sacrifice in multiple ways that connote animal sacrifice. He has also established Iphigenia as an origin myth for structural violence against women and animals. The Iphigenia myth not only reveals what Lucretius believes to be the essence of *religio*, it also does so by characterizing religion's violence against women and animals as foundational for Roman politics. The idea that religious violence is part of the deep structure of Roman politics is suggested most clearly by Lucretius' anachronistic use of *civis* ('citizens', line 91). Most translators, precisely in order to avoid anachronism, translate *civis* here as 'countrymen', as I have done above. But there was no such thing as 'citizens' at the time Iphigenia's sacrifice is represented as taking place. The term *civis* is, therefore, not just anachronistic, but strikingly so. I would claim that this anachronism is intentional on Lucretius' part. It is meant to mark the religious violence of Iphigenia's sacrifice as a genealogical origin revealing the essence of *religio*. By placing 'citizens' back into a foundation myth about male religious violence against a woman sacrificed like an animal, Lucretius

---

23   For the general background to sacrifice as gift-exchange with the gods, cf. Scheid 2007, pp. 149-50.

seems to be suggesting that the interconnections between a religion, politics, women, and animals is far from ancient history. Rather, it is part of the sedimentary structure of Roman political life.[24] As classicist Alessandro Schiesaro has pointed out, Lucretius is criticizing not so much religion in general as religion's social and political function in specific. Schiesaro reminds us that for Polybius, the Greek historian who narrated the rise of Rome, religion had always been considered by Romans to be 'the foundation of Roman greatness (6.56)'. Schiesaro, for his part, also points out the political function of the word 'king' in this passage (*regem*, line 94), which he says, would have reminded Romans of this period about their own troubling political origins in religious kingship. *Rex* was undoubtedly one of the most hated words in the Latin language. The sacrifice of Iphigenia, therefore, Schiesaro claims, was 'couched in the language of Roman officialdom' in order to reveal for just how long patriarchical religion had been a tool for men in power.[25]

Lucretius concludes the sacrifice of Iphigenia with a curious supplement, i.e. a statement that simultaneously both *adds* to the passage and *completes* it. This is the most famous line from *De rerum natura*: 'So powerful was religion in persuading evils' (*tantum religio potuit suadere malorum*, line 101). Imposing a statement that is both an addition and also simultaneously a completion introduces a certain tension. A supplement (an addition from the outside) and a conclusion (a reiteration of the inside, i.e. of the content) are not necessarily the same thing. And yet, such 'supplementary logic' is not a concept that began with Derrida, or Rousseau.[26] In the ancient rhetorical handbooks, there exists a similar rhetorical figure called an *epiphonema* (ἐπιφώνημα, 'additional utterance'). Here, I would agree with Peta Fowler that the sentence 'So powerful was religion in persuading

---

24   Munro wonders apropos Lucretius' use of *civis*, 'What did Lucr. think of the fate of his own countrymen the Decii?' Cf. Munro, p. 128.

25   Cf. Alessandro Schiesaro, 'Lucretius and Roman *politics* and *history*', in *The Cambridge Companion to Lucretius*, ed. by Stuart Gillespie and Philip Hardie (Cambridge: Cambridge University Press, 2007), p. 52.

26   Most famously, Jacques Derrida introduced the 'logic of supplementarity', or, in his famous phrase, '...That Dangerous Supplement...' in his readings of both Jean-Jacques Rousseau and Claude Levi-Strauss in *Of Grammatology*, trans. Gayatri Spivak (Baltimore: Johns Hopkins University Press, [1976] 1997), pp. 141-165, especially 144-5. Cf. also *Writing and Difference* (Chicago: University of Chicago Press, 1978), p. 289; and *Given Time I: Counterfeit Money* (Chicago: University of Chicago Press, 1992), pp. 66-7.

evils' is an *epiphonema*.[27] It not only *summarizes* the myth, in that it reveals the essence of religion to be its power for evil (literally, 'what evil it can do': *potuit*, line 101). It also *adds*, or introduces from the outside, the additional idea that religious violence is 'sweet' (*suadere*, 'to persuade' or 'to make sweet'). Sweetness was not an idea contained in the Iphigenia passage, but is imposed by Lucretius from the outside and supplemented at the end. This supplementarity of the 'sweetness' of religious violence raises numerous questions. One of the most pressing of which, for a poet like Lucretius, is how to present a poetry of vitalism without violence. Though Lucretius means to condemn the idea that masculine religious violence is sweet, or persuasive, nonetheless, the poetry itself in the passage is sweet in the sense of being a virtuoso performance. And yet, that sweetness is in tension with the message of the passage. As an Epicurean hedonist, Lucretius believed sweetness to be a value in poetry. The sacrifice of Iphigenia, therefore, is sweet and persuasive, but not in the way that the perpetrators of the sacrifice, the male priests and army leaders, would recognize. In Lucretius' view, their sense of what is sweet and persuasive has been completely perverted by false views about nature and religion.

Human Soul, Animal Body: the Influence of Empedocles

So far this chapter has focused on the absent animal referents in the sacrifice of Iphigenia, and how those referents are part of Lucretius' intersectional critique of patriarchical religion. It is necessary, however, to take a detour at this point to discuss the most direct influence on Lucretius' description of Iphigenia in Book One of *De rerum natura*, as well as the most direct influence on his description of the sacrificed calf in Book Two, described in the following section.

---

27 For *epiphonema* as an ancient rhetorical device, cf. Peta Fowler, 'Lucretian Conclusions', in *Oxford Readings in Classical Studies: Lucretius*, ed. by Monica Gale (Oxford: Oxford University Press, [2007] 2011), pp. 204-5 and note 13. David Sedley has noted Lucretius' tendency to introduce new ideas into the conclusions of his arguments, a fact which suggests that there is more to do with supplementary logic in Lucretius. Cf. Sedley's reading of *divinitus* at *De rerum natura* 1.149-58, in David Sedley, *Lucretius and the Transformation of Greek Wisdom* (Cambridge: Cambridge University Press, [1998] 2003), pp. 198-9.

Scholars agree that Lucretius' description of Iphigenia is most directly in dialogue with a fragment that has survived from the Greek philosophical poet and Pythagorean-vegetarian Empedocles, who lived about four hundred years before Lucretius. While Lucretius does not share Empedocles' Pythagorean belief in the transmigration of human souls into animal bodies, he does share Empedocles' horror of sacrifice.[28] The fragment of Empedocles printed here, scholars agree, was the most proximate source for Lucretius' description of the sacrifice of both Iphigenia and of the calf described in the next section.[29] In the fragment here, Empedocles describes a father who has unknowingly sacrificed and eaten his son, because his son appears to him as an animal, having transmigrated into an animal's body. Empedocles means to warn us about the dangers of eating the flesh of any living being, human and nonhuman alike.

μορφὴν δ' ἀλλάξαντα πατὴρ φίλον υἱὸν ἀείρας
σφάξει ἐπευχόμενος μέγα νήπιος †οἱ δὲ πορεῦνται†
λισσόμενον θύοντες· †ὁ δ' αὖ νήκουστος† ὁμοκλέων
σφάξας ἐν μεγάροισι κακὴν ἀλεγύνατο δαῖτα.
ὡς δ' αὔτως πατέρ' υἱὸς ἑλὼν καὶ μητέρα παῖδες
θυμὸν ἀπορραίσαντε φίλας κατὰ σάρκας ἔδουσιν.

The father will lift up his dear son in changed form and cut his throat, the great fool, while he prays. The attendants bring in the pleading victim; but the   father is deaf to its cries and, having carried out the sacrifice, prepares an evil feast in his halls. In the same way the son seizes his father and children [seize] their mother, and having ripped away her life they eat her loving flesh.[30]

In both the passage about Iphigenia above and the fragment from Empedocles here there are numerous similarities. In both, a father

---

28   Cf. Myrto Garani, *Empedocles Redivivus: Poetry and Analogy in Lucretius* (London: Routledge [2007] 2012), p. 246, note 186.

29   See note 35 below for discussion about the influences behind Lucretius' description of Iphigenia and the sacrificed calf.

30   For the text of Empedocles reproduced here, and for general commentary on the fragment, cf. M.R. Wright, *Empedocles: the Extant Fragments* (Bristol: Bristol Classical Press, [1981] 2010), p. 145 for the text and pp. 286-7 for the commentary; as well as Brad Inwood, *The Poem of Empedocles: A Text and Translation* (Toronto: University of Toronto Press, 2001), p. 271 for the text and p. 289 for textual notes.

sacrifices a child, described as an animal. In both, the animal is really a human. But in Empedocles the human child retains his human soul while still inhabiting an animal body, retaining his full capacity for human consciousness and feeling. In Lucretius, Iphigenia is fully human, and described as an animal, but her animality is described through use of the absent referent. In Empedocles, the situation is almost the reverse. The victim has the outward form of an animal, but remains human inside since he retains his human soul. The important point is that, whereas Lucretius uses metaphor to make intersectional connections between animals and humans, for Empedocles human and animal souls can actually inhabit each other's bodies. As a Pythagorean sympathizer, Empedocles believed that animals can have human souls and humans can have animal souls. This is why Empedocles warns his readers against eating or abusing animals. They may be, or have once been, human.

There are further similarities at the level of language. In both Lucretius and Empedocles, the victims are 'raised up' – *sublata* in Lucretius (line 95 above) and *aeiras* in Empedocles here (ἀείρας, line 1). Raising or lifting the victim was a practice typical when sacrificing small animals. The raising up of the victims in Lucretius and Empedocles, therefore, makes sense, because the animals referred to in both are young animals (the youthful Iphigenia and the child/animal here).[31] In both Lucretius and Empedocles, the father also does the killing with his own hand. But in Lucretius, Iphigenia is said to be mute with dread (*muta metu*, line 92 above). Like an animal, she cannot speak on account of her fear. In the fragment from Empedocles here, there is a notable contrast. Empedocles shows the soul of the child struggling to break through the limitations of its animal body and its animal voice by crying out.[32] And yet, despite the humanity of the animal's cries, the father – the 'great fool' (*mega nepios*/μέγα νήπιος, line 2), as Empedocles calls him – is deaf (another meaning of *nepios*/νήπιος) to his son's cries.[33]

---

31 Adult animals, however, were considered more suitable for public worship, which is yet another way in which both the sacrifice of Iphigenia, of the child here, and of the calf below would be considered ritual failures (cf. Scheid 2011, p. 264). It is worth noting that in Aeschylus' play *Agamemnon*, Iphigenia is also said to have been raised up (*Agamemnon*, lines 232-5).

32 Cf. λισσόμενον, 'beseeching'/'praying': the child's human soul is 'beseeching' his father, almost as in prayer, from within his animal body.

33 There is a multi-layered pun in the Greek phrase 'great fool' (μέγα νήπιος), in that νήπιος primarily means 'infant' (in the etymological sense of one who

Empedocles' point is that when you eat meat you do not know whose soul you are eating. This is because Empedocles believes that souls change their outward shape (*morphen allaxanta*/μορφὴν ἀλλάξαντα), while at the same time retaining their identity. In Lucretius, however, human and animal souls do not change their outward shape. As a materialist atomist, Lucretius believes that you are your physical shape. Nonetheless, humans and animals in Lucretius are portrayed as suffering nearly the same mental and physical pain, as we will see in the next section.

What needs to be emphasized is that, while for Empedocles harming animals is problematic because you can never really be sure what kind of soul you are harming, for Lucretius, the outward appearance of an animal is a reliable indicator of its identity. The Epicureans were notorious for believing that all appearances are true. At the same time, Lucretius seems to think that the mental capacities of humans and animals are so similar that it is not merely metaphorical for animals to stand in as absent referents for humans, and humans for animals. This is because, for Lucretius, when it comes to suffering, the interchangeability of human and animal mind is at its most fluid and sympathetic.

Making the Absent Referent Present: Mother Cow and Her Calf

The intertextual connections between Empedocles and Lucretius are both direct and indirect. It is a well-established fact that Lucretius read Empedocles directly.[34] But Lucretius also knew Aeschylus' portrait of

---

cannot speak, cf. the Latin *in-fans*). By extension, νήπιος also means 'fool', in the sense that speechlessness suggests childlike foolishness. This is ironic, first, because the young animal being sacrificed really is a child. It is a young animal, whereas the father is an mature human. It is also ironic in the sense that the child cannot speak on account of his animal body: he is νήπιος in the sense of 'speechless'. Finally, it is also ironic in a way that focuses on the father: the father is a 'fool', a νήπιος, in the sense that he is deaf to the utterances of his son, i.e. he is 'speechless' in the sense of not hearing the speech of his son. This third layer of irony is possible because νήπιος in Greek, like *in-fans* in Latin, means most literally 'not-word'/'not-utterance' (νη- is a negative prefix, and ηπιος comes from ἔπος, which means 'word'/'utterance'). Most obviously, of course, the father is a νήπιος for sacrificing his son in the first place.

34   Cf. Myrto Garani, *Empedocles Redivivus: Poetry and Analogy in Lucretius* (London: Routledge, [2007] 2012), the most recent and extensive treatment of Empedocles' influence on Lucretius.

Iphigenia in *Agamemnon*, which itself drew on the same Empedocles fragment. The Iphigenia myth was, of course, also well known in classical culture in general. The Empedocles fragment, therefore, stands behind the Iphigenia passage above both directly, through Lucretius' reading of Empedocles, and indirectly, through Lucretius' reading of Aeschylus', who also read Empedocles. And as we will see below, the Empedocles fragment was also an intertext for Lucretius' description of the sacrifice of the calf. The sacrifice of Iphigenia above and the sacrifice of the calf below are, therefore, not just linked thematically, or hermeneutically – i.e. by reading them closely together. They are also linked by their sources, having, as a common source, the Empedocles fragment described above. Empedocles' horror of animal sacrifice, therefore, can be heard speaking through the Iphigenia passage above, the myth about the father who sacrifices his child just discussed, and the sacrifice of calf below.[35]

Establishing this intertextual connection is important because it helps to show the multiple lines (hermeneutic, thematic, intertextual) that link the absent referent to animality in the sacrifice of Iphigenia and make it present again in the sacrifice of the calf below. These various connections also play a role in connecting the human-like grief of the mother cow below with the grief of Iphigenia's own mother, Clytaemnestra. Clytaemnestra is absent from Lucretius' telling of the Iphigenia myth above. The mother cow and her grief, however, make the absence of Clytaemnestra's grief present again in the mother cow's search for her murdered calf.

Here, Lucretius makes the absent referent present again by turning the politics of metaphor on its head. For once, human tragedy is forced to play the role of absent referent in the tragedy of animals, and human suffering takes a backseat to animal suffering. In the passage below, Lucretius

---

35   Cf. David Furley, 'Variations on Themes from Empedocles in Lucretius' Proem', *Bulletin of the Institute of Classical Studies*, 17 (1970), pp. 55-64 (p. 62); Monica Gale [2000] 2006, pp. 103-5; Gale [1994] 2007, p. 72. Furley notes that the fragment from Empedocles was itself a source text for the sacrifice of Iphigenia in Aeschylus' play *Agamemnon*. David Sedley adds that the portrait of Iphigenia from Aeschylus' *Agamemnon* and Euripides *Iphigenia at Aulis* and *Iphigenia at Tauris* were, unlike the Empedocles fragment, not direct sources for Lucretius (Cf. Sedley [1998] 2003, p.30). Hence, my focus in this chapter on the influence of the Empedocles, and not Greek tragedy.

depicts a mother cow searching for her lost calf who unbeknownst to her
has just been pointlessly slaughtered on an altar to the gods.

quorum unum quidvis generatim sumere perge:	347
invenies tamen inter se differe figuris.	348
nec ratione alia proles cognoscere matrem	349
nec mater posset prolem – quod posse videmus	350
nec minus atque homines inter se nota cluere.	351
nam saepe ante deum vitulus delubra decora	352
turicremas propter mactatus concidit aras,	353
sanguinis expirans calidum de pectore flumen;	354
at mater viridis saltus orbata peragrans	355
quaerit humi pedibus vestigia pressa bisulcis,	356
omnia convisens oculis loca si queat usquam	357
conspicere amissum fetum, completque querellis	358
frondiferum nemus adsistens et crebra revisit	359
ad stabulum desiderio perfixa iuvenci;	360
nec tenerae salices atque herbae rore vigentes	361
fluminaque illa queunt summis labentia ripis	362
oblectare animum subitamque avertere curam,	363
nec vitulorum aliae species per pabula laeta	364
derivare queunt animum curaque levare;	365
usque adeo quiddam proprium notumque requirit.	366

Go on, take any animal kind (in nature): you will find, nevertheless,
that they differ from each other in figure. In no other way could a child
recognize its mother, nor a mother her child-which we see they can do,
and are no less known to each other than humans are. For often in front
of ornate temples of the gods a calf falls sacrificed beside the altars of
burning incense, breathing out a hot river of blood from its chest; but the
abandoned mother, wandering through the life-green valley, searches
the ground for tracks made by cloven hooves, surveying every region
in the hope of seeing her lost offspring again, and coming to a stop
at a leaf-bearing grove fills it with lamentations, and frequently going
back to look in the stable, stabbed with longing for her calf; neither can
young willow-shoots nor grasses enlivened with dew nor those rivers
flowing atop their banks delight her mind and remove the sudden pain,
nor can the sight of other calves in happy meadows lead her mind into
other channels and lighten her pain: so persistently does she seek what
is her own and known to her.

Here we get a 'cow's eye view', as classicist David West puts it, of the 'leaf-bearing groves', 'young willow-shoots', 'grasses enlivened with dew', and 'rivers flowing atop their banks', as the cow searches the landscape for her calf.[36] Lucretius, it is worth remembering, also portrayed the sacrifice of Iphigenia from her perspective. She was blindfolded and mute with dread, capable only of animal sensation, feeling for the dangers around her, and expressing her emotions through her trembling body. Both scenes, therefore, valorize the emotional perspectives of the victims.[37] For both Iphigenia and for the mother cow the focus is on what male-dominated *religio* looks like through the eyes of bare life, that is, through the eyes of the cast-out and killable.[38] The narrative voice also slips into the thoughts and merges with feelings of the victims. Though told for a third-person persepective, the narrative voice is hardly objective or distanced from the suffering of the victims, the opposite of what is usually thought to be characteristic of epic.[39]

Lucretius humanizes the grieving cow by calling her 'mother' (*mater*, lines 349/350) and 'abandoned mother' (*orbata mater*, line 355). Lucretius also calls her calf 'child' (*prolem*, line 350). While Lucretius does not share Empedocles' belief in the transmigration of souls, he does represent the emotional and cognitive responses of the mother cow as what a human mother's soul would experience in such a situation. Furthermore, Lucretius equates the bond between human mothers and children and animal mothers and children, when he says that animals are 'no less *known* to each other than humans are' (*nec minus atque homines inter se* nota *cluere*, line 350). Lucretius further specifies this relationship when he says that 'you will find that each particular animal differs from each in figures' (*in figuris*, line 348). That is, while animal kinds are generally similar in their material make-up, each individual has a particular *figura*, or composite shape, that

---

36   David West, 'Two Plagues: Vergil, Georgics 3.478-566 and Lucretius 6.1090-1286', in *Creative Imitation in Latin Literature*, ed. by D.A. West and A.J. Woodman (Cambridge: Cambridge University Press, 1979) pp. 71-88.

37   Cf. Morrison, pp. 214-5; E.J. Kenney, *Lucretius: De rerum natura Book III* (Cambridge: Cambridge University Press, 1971), p. 16.

38   For 'bare life' see note 17 above.

39   While written from the third-person perspective, the descriptions are, nonetheless, heavily colored by the perspectives and emotions of the victims. Some would call this focalization, but its intensity suggests to me something closer to 'free indirect discourse', that is, closer to narrative discourse that slips in and out of the minds and feelings – and not just of the points-of-view – of the victims.

its mother – human or animal – knows as its own 'in no other way'
(*nec ratione alia*, line 349). Here, Lucretius is talking about animal
knowledge, but of a specific kind, what we might call 'acquaintance
knowledge'. This is the knowledge that any mother has of her child,
insofar as that knowledge comes about through proximity, physical
interaction, acquaintance, and most importantly, in Lucretius' view,
through the mother having 'fashioned' (*fingere*) the 'figure' (*figura*) of
her child in her womb. *Figura* and *fingere* are etymologically related.
Thus, the 'composite shape' of the child, its *figura*, is particularly
well known to animal mothers, because they are the ones who have
'fashioned' (*fingere*) its 'figure' (*figura*) in the womb. The maker knows
what she has made.

Lucretius' choice of *figura* is particularly resonant. As Eric Auerbach
claimed: 'there is no doubt that of all the authors I have studied in
connection with *figura*, it was Lucretius who made the most brilliant,
though not the most historically important contribution'.[40] In Auerbach's
view, *figura* in Lucretius does not mean merely 'outward shape' or
'plastic form'. Rather, *figura* translates the Greek word *typos* (τύπος),
which means 'imprint', 'seal', or 'model', and therefore indicates
a recognizable physical stamp or seal that, I would add, any animal
mother knows in her child, since she made it. Animal mothers recognize
themselves as the physical models of the copies of themselves, their
child. This is not knowledge of the personality of the child. Nor is it
the ability to have the proposition-based thought, i.e. to be able to say
to oneself in language: 'this is my child'. Rather, an animal knowledge
based on figurality – that is, based on the ability to recognize the *figura*
of one's child – would represent a kind of pre-propositional, sensation-
and acquaintance-based form of knowledge that all animal mothers
share. It is knowledge of biological model to biological copy. Auerbach
saw this connection when discussing how *figura* means more than
forma ('form') in Lucretius:

> [*Figura*] may be best noted in the passage dealing with the
> resemblance of children to their parents, the mixture of seeds, and
> heredity; with children who are *utriusque figurae* ('of both *figurae*'),
> resembling both father and mother, and who often reflect *proavorum*
> *figuras* ('the *figurae* of their ancestors'), and so on: *inde Venus varias*

---

40   Eric Auerbach, *Scenes from the Drama of European Literature: Six Essays*
      (New York: Meridian Books, 1959), pp. 17-18.

producit sorte figuras ('thence Venus brought forth diverse *figurae* in turn') (4,1223). Here we see that only *figura* could serve for this play on model and copy.[41]

Therefore, when Lucretius says that 'each particular animal differs from each in figure (*in figuris*)', he is using a term that is not necessarily restricted to humans, and that is used not only to signify the outward shape of the animal but also the biological relationship of model (the mother) to the biological copy she has fashioned (her child). Such a biological 'stamp' (another meaning of *figura*) is knowable through the animal knowledge that belongs to every human and animal mother.

Like Iphigenia, the calf's youth and life are cut short through the pointless cruelty of religion. The calf dies while 'breathing out a hot river of blood' *sanguinis expirans calidum de pectore flumen* (line 354).[42] The mother cow is also 'pierced through the heart with longing' for her child (*perfixa desiderio*, line 360), as if in sympathy with the pain her child felt when wounded in the chest (*de pectore*, line 354). Lucretius contrasts this vivid violence with the vitalism of

---

41   Auerbach, p. 16. Also note the genetic connotations of *figura* – as opposed to shape (*forma*), the latter of which is associated with spatiality – at *De rerum natura* 2.333-5: *Nunc age iam deinceps cunctarum exordia rerum/qualia sint, et quam longe distantia formis/percipe,* multigenis *quam sint variata* figuris 'Come now, and learn next what kinds there are of the beginnings of all things, and perceive how far distant they are in shape, how varied they are in *multigenic figures*'. Cf. P.H. Schrijvers, *Horror Ac Divina Voluptas: Etudes sur la poetique et la poesie de Lucrece* (Amsterdam: A.M. Hakkert, 1970), p. 223; and, more generally, Brooke Holmes and William Henry Shearin on the importance of Auerbach's reading of *figura* in Lucretius, who note that 'Lucretius himself, if not Epicureanism more generally, makes a substantial contribution to our understanding of what *figura* (and 'figural') may mean'. Cf. Holmes and Shearin, 'Introduction: Swerves, Events, and Unexpected Effects', in *Dynamic Reading: Studies in the Reception of Epicureanism*, ed. by Brooke Holmes and William Henry Shearin (Oxford: Oxford University Press, 2012), p. 11, note 20.

42   Cf. Charles Segal, 'Delubra decora: Lucr. II.352-66', *Latomus*, 29 (1970), 104-18. The blood of small animal victims was allowed to blacken the altar or drip into the hearth or sacrificial pit, whereas a bowl was used to catch the blood of larger victims. The point in both cases was that no blood should fall on the earth (cf. Bremmer, 137). In the case of the sacrifice of the calf in Lucretius, even though the victim is small, the excess of blood may indicate that the sacrifice had gone awry. For surely some of the 'river of blood' (*sanguinis…flumen*) would have fallen on the earth. That is, like the sacrifice of Iphigenia, this too is a ritual failure.

the natural landscape: the lush, flowing paradise, or *locus amoenus*, that the mother cow has to pass through grieving to find her child. Despite its attractions, however, Lucretius tells us that the landscape did not 'divert her mind into other channels' (*derivare animum*, line 365). Lucretius is here activating the latent metaphor in *derivare* ('to turn the course of a river'), contrasting the flows of the cow's mind with the flows of the landscape. The mother cow, nonetheless, keeps the flow of her concentration directed towards her child.

None of the aspects of the earthly paradise that she traverses – all of which are described in language that echoes the absence of the youth and life lost in her child – are able (*queunt*, lines 362/365) to delight her mind and take away the pain.[43] The landscape cannot turn away her psychological pain (*avertere curam*, line 363). And yet, the landscape that we see through her eyes fails to calm her not because she is too unintelligent to enjoy it, but because, as Lucretius says at the end of the passage: 'so persistently does she seek what is her own and known to her' (*usque adeo quiddam proprium notumque requirit*, line 366). Her persistence for what is 'her own' (*proprium*) and 'known to her' (*notum*), and not mere instinct, keeps the landscape from diverting the flow of her mind through the allurements of pleasure.[44] Instead, she *chooses* to pursue her child, despite the pleasures all around her that would ensnare an animal acting out of mere instinct or only temporary awareness.

I want to end this chapter by returning to Carol Adams' notion of the absent referent. Lucretius has made the absent referent in the sacrifice of Iphigenia vividly present in the sacrifice of the calf. As Adams puts it, the absent referent: 'was what enabled the interweaving of the oppression of women and animals. Behind every meal of meat is an absence: the death of the animal whose place the meat takes. The

---

43   *Ob-lectare animum* (line 363) means, literally, to 'delight-away her mind'. *Oblectare* may also pun on *lac*, 'milk'. Consider in this regard line 370, which comes just after the passage quoted: '*each* [young animal] usually runs down to *its own* udder of *milk*' (ad sua quisque *fere decurrunt ubera* lactis).

44   Lucretius' depiction of animal knowledge as acquaintance knowledge (*proprium notumque*, line 366) comes close to Heidegger's notion that animals have knowledge in terms of 'property' (*Eigentum*). Heidegger seems to believe that animal knowledge amounts to being able to recognize what is proper to them, what is their own, as he says, 'The way and manner in which the animal is proper to itself is not that of personality, not reflection or consciousness, but simply its proper being (*Eigentum*)'. Cf. *The Fundamental Concepts of Metaphysics: World, Finitude, Solitude*, trans. by William McNeill and Nicholas Walker (Bloomington: Indiana University Press, 1995), p. 233.

'absent referent'', Adams continues, 'is that which separates the meat eater from the animal and the animal from the end product. The function of the absent referent is to keep our 'meat' separated from any idea that she or he was once an animal...to keep something from being seen as having been someone. Once the existence of meat is disconnected from the existence of an animal who was killed... [it becomes] instead a free-floating image, used often to reflect women's status as well as animals'. Animals are the absent referent in the act of meat eating; they also become the absent referent in images of women butchered, fragmented, or consumable'.[45]

Animals, Adams argues, are used as metaphors to heighten scenes of female butchery: the animal's fate is transferred to the fate of the human victim. The danger of such metaphors, however, is that, as Adams noted in *The Sexual Politics of Meat*, 'the original meaning of the animal's fate is absorbed into a human-centered hierarchy'. The real suffering of animals becomes an instrument in a hierarchy of meanings that prioritizes human suffering. But, as I hope to have shown, by making the absent animal referent in the sacrifice of Iphigenia present in the sacrifice of the calf, Lucretius not only equalizes (in a non-, or at least minimally, speciesist way) the suffering of both women and feminized animals under male-dominated *religio*, but also valorizes the animal's point of view.

This valorization comes about partly because, in both the sacrifice of Iphigenia and of the calf, Lucretius is adapting Empedocles' idea that sacrificial victims may have human souls. Lucretius is, therefore, not setting up a mere analogy between the suffering of women and animals under *religio*, but rather is making an intersectional argument about the ideological structure that binds female suffering across species lines. Iphigenia and the mother cow are not merely analogous in their pain and misfortune but nearly identical in their mental and physical suffering, because of the interconnectedness of the structural violence established under male-dominated *religio*.

The sacrifice of women and animals in Lucretius, therefore, does not present male religious violence against animals as a merely an additive or additional form of injustice: animal suffering is not merely analogous to the suffering of women. Rather, Lucretius presents the suffering of women and animals under patriarchy as thoroughly interlocking, all the way down to the figural bonds that are broken between human

---

45   Adams 1990, p. 13.

and animal parents and children by religious violence. If a goal of ecofeminism is to reveal the structural interconnections of male violence against women, (feminized) men, and (feminized) animals, as I quoted from Carol Adams at the beginning of this chapter, then Lucretius has not only tried to identify the origins of that system in Roman culture, but has also given us original myths to critique it.

Bibliography

Adams, Carol, *The Sexual Politics of Meat: A Feminist-Vegetarian Critical Theory* (London: Bloomsbury, 2010 [1990]).

Adams, Carol, 'Woman-Battering and Harm to Animals', in *Animals and Women: Feminist Theoretical Explorations*, ed. by Carol Adams and Josephine Donovan (Durham: DukeUniversity Press, 1995) pp. 55-84.

Agamben, Giorgio, *Homo Sacer: Sovereign Power and Bare Life*, trans. by Daniel Heller-Roazen (Stanford University Press, 1998).

Auerbach, Erich, *Scenes from the Drama of European Literature: Six Essays* (New York: Meridian Books, 1959).

Bailey, Cyril, *T. Lucreti Cari De rerum natura*, 3 vols. (Oxford: Oxford University Press, [1947] 2001).

Bakhtin, Mikhail, 'Discourse in the Novel', in *The Dialogic Imagination: Four Essays* (Austin: Texas University Press, 1981), pp. 259-422.

Bremmer, Jan, 'Greek Normative Animal Sacrifice', in *A Companion to Greek Religion*, ed. by Daniel Ogden (Wiley-Blackwell, [2007] 2010) pp.132-144.

Burkert, Walter, *Greek Religion*, trans. by John Raffan (Cambridge: Harvard University Press, [1985] 2000).

Deleuze, Gilles, 'Spinoza and the Three 'Ethics'', in *Essays Critical and Clinical*, trans. by Daniel W. Smith and Michael A. Greco (Minneapolis: University of Minnesota Press, 1997).

Derrida, Jacques, *The Animal That Therefore I am*, trans. by David Wills (New York: Fordham University Press, 2008).

Fowler, Peta, 'Lucretian Conclusions', in *Oxford Readings in Classical Studies: Lucretius*, ed. by Monica Gale (Oxford: Oxford University Press, [2007] 2011).

Furley, David, 'Variations on Themes from Empedocles in Lucretius' Proem', *Bulletin of the Institute of Classical Studies*, 17 (1970), 55-64.

Gale, Monica, *Virgil on the Nature of Things* (Cambridge: Cambridge University Press, [2000] 2006).

Gale, Monica, *Myth and Poetry in Lucretius* (Cambridge: Cambridge University Press, [1994] 2007).

Garani, Myrto, *Empedocles Redivivus: Poetry and Analogy in Lucretius* (London: Routledge, [2007] 2012).

Heidegger, Martin, *The Fundamental Concepts of Metaphysics: World,*

Finitude, Solitude, trans. by William McNeill and Nicholas Walker (Bloomington: Indiana University Press, 1995).

Holmes, Brooke and Wiliam Henry Shearin, 'Introduction: Swerves, Events, and Unexpected Effects', in *Dynamic Reading: Studies in the Reception of Epicureanism*, ed. by Brooke Holmes and William Henry Shearin (Oxford: Oxford University Press, 2012).

Inwood, Brad, *The Poem of Empedocles: A Text and Translation* (Toronto: University of Toronto Press, 2001).

Kenney, E.J., *Lucretius: De rerum natura Book III* (Cambridge: Cambridge University Press, 1971).

Kristeva, Julia, *New Maladies of the Soul*, trans. by Ross Guberman (New York: Columbia University Press, 1995).

Leonard, William Elley and Stanely Barney Smith, *T. Lucreti Cari De rerum natura libri sex* (Madison: University of Wisconsin Press, [1942] 1968).

Morrison, A.D., '*Nil igitur mors est ad nos*? Iphianassa, the Athenian Plague, and Epicurean Views of Death', in *Lucretius: Poetry, Philosophy, Science*, ed. by Daryn Lehoux, A.D. Morrison, Alison Sharrock (Oxford: Oxford University Press, 2013).

Munro, Hugh Andrew Johnstone, *Titi Lucreti Cari De Rerum Natura Libri Sex, with a Translation and Notes* (Cambridge: Cambridge University Press, [1846] 2010).

Scheid, John, *An Introduction to Roman Religion*, trans. by Janet Lloyd (Bloomington: Indiana University Press, 2003).

Scheid, John, 'Sacrifices for the Gods and Ancestors', in *A Companion to Roman Religion*, ed. by Jorg Rupke (Wiley-Blackwell, [2007] 2011) pp. 263-71.

Schiesaro, Alessandro, 'Lucretius and Roman politics and history', in *The Cambridge Companion to Lucretius*, ed. by Stuart Gillespie and Philip Hardie (Cambridge: Cambridge University Press, 2007).

Schrijvers, P.H., *Horror Ac Divina Voluptas: Etudes sur la poetique et la poesie de Lucrece* (Amsterdam: A.M. Hakkert, 1970).

Sedley, David, *Lucretius and the Transformation of Greek Wisdom* (Cambridge: Cambridge University Press, [1998] 2003).

Segal, Charles, '*Delubra decora*: Lucr. II.352-66', *Latomus*, 29 (1970), 104-18.

West, David, 'Two Plagues: Vergil, Georgics 3.478-566 and Lucretius 6.1090-1286', in *Creative Imitation in Latin Literature*, ed. by D.A. West and A.J. Woodman (Cambridge: Cambridge University Press, 1979) pp. 71-88.

Wolfe, Cary, *Before the Law: Humans and Other Animals in a Biopolitical Frame* (Chicago: University of Chicago Press, 2013).

Wright, M.R., *Empedocles: the Extant Fragments* (Bristol: Bristol Classical Press, [1981] 2010).

AMADEUSZ JUST

SCHOPENHAUER AND DIAMOND ON ANIMAL ETHICS

One of the most popular nineteenth-century arguments against cruelty to animals, an argument present also in Immanuel Kant's writings,[1] was that cruelty to animals may lead to cruelty to people. Arthur Schopenhauer explicitly expressed his dissatisfaction with this argumentation, which he could find presented by newly established animal welfare societies (the earliest European welfare organisation, created in England, dates back to 1824, while the first German association was established in 1837), in the second volume of *Parerga and Paralipomena*: 'In their exhortations the societies for the protection of animals are for ever using the bad argument that cruelty to animals leads to cruelty to human beings, as though man were a direct object of moral duty, the animal being merely indirect, in itself a mere thing. 'For shame!''.[2] A year before the publication of these words, in 1850, Schopenhauer sent a copy of his book *Die beiden Grundprobleme der Ethik* including discussion of his morality of compassion to the chairman of the *Müncher Tierschutzverein* (Münchner society for the prevention of cruelty to animals) – Ignaz Perner. Not only Perner gratefully appreciated the gift, but also it was the beginning of Schopenhauer's engagement in animal welfare organisations.[3] Although it is said that Perner popularized Schopenhauer's ethics in his talks and pamphlets, the morality of compassion does not contribute to the current debates on animals. For more than four decades the most influential ideas opposing cruelty to animals and grounding animal liberation movement are Peter Singer's utilitarianism and

---

1   See: Alexander Broadie, Elizabeth M. Pybus, 'Kant's Treatment of Animals', Philosophy, vol. 49, no. 190 (Oct., 1974), 375-383.

2   Arthur Schopenhauer, *Parerga and Paralipomena*, vol. 2, trans. by Eric F. J. Payne (Oxford: Clarendon Press, 1974; repr. 2000), p. 372.

3   For Schopenhauer's engagement with European welfare societies see: Monica Libell, *Morality Beyond Humanity. Schopenhauer, Grysanowski and Schweizer on Animal Ethics* (LAP: Saarbrücken, 2010), pp. 140-146.

Tom Regan's animal rights theory. Both of these views proceed with an *argument from common capacities*,[4] which claims that to remain consistent we should treat all beings – regardless of their species – which possess the appropriate capacities, equally, unless our moral attitude is to be marked by *speciesism*. For Singer these appropriate capacities which ground our equal moral considerations consist in the capacity of suffering and enjoyment. Regan's version of the *argument from common capacities* slightly differs, as he argues for rights of certain animals on the basis of animals' subjectivity, where feelings of pleasure and pain are only one of its features. The Reagan-Singer's approach faced Cora Diamond's opposition. Diamond's paper *Eating Meat, Eating People* is a direct response to this sort of *arguments from common capacities*. She claims, in a controversial phrase, that: 'We cannot point and say, 'This *thing* (whatever concepts it may fall under) is at any rate capable of suffering, so we ought not to make it suffer.''[5]

Schopenhauer, rejecting the *indirect claim view* that cruelty to animals may indirectly lead to cruelty to human beings, and Diamond, opposing Regan-Singer's argument from common capacities, are not suggesting that there is nothing wrong with animal mistreatment. On the contrary, both of them are keenly interested in the ethical issue of how human beings treat animals and precisely from this interest results their concern and dissatisfaction with those sort of arguments, because 'they – as Diamond put it – contain fundamental confusions about moral relations between people and people *and* between people and animals'.[6] In this paper I would like to discuss this *fundamental confusions* as Schopenhauer and Diamond understand it and take a closer look at what they propose instead, i.e. what are the features of a genuine description of the moral relations between people and animals.

---

4   This argument is usually described as the argument from marginal cases as it turns on an analogy between animals and mentally disabled/ limited/'marginal' human beings. Nonetheless, in this paper I am going to follow Alice Crary's advice and use her neutral label *argument from common capacities*.

5   Cora Diamond, 'Eating Meat, Eating People', *The Realistic Spirit. Wittgenstein, Philosophy, and the Mind* (Cambridge, London: A Bradford Book, The MIT Press, 1991), p. 325.

6   Ibid., p. 319.

I

Before I move to the main part of my paper I have to make one general remark. A juxtaposition of Arthur Schopenhauer's theory of morality with Cora Diamond's ethical views may seem merely an intellectual play. The nineteenth-century German idealist, best known for his metaphysics of the Will and pessimistic *Weltanschauung*, may seem to have nothing to do with the contemporary American thinker known for her original interpretations of Ludwig Wittgenstein's philosophy. Therefore, it may seem that I am going to claim that Diamond's insights are already present in Schopenhauer's philosophy, if only we read it properly. If I were to do so, indeed, it would be a futile intellectual effort. But there is one crucial reason which justifies comparing Schopenhauer with Diamond. In 1963 Alfred J. Ayer noted that 'admirers of Wittgenstein may be surprised to discover the extent to which his thought was influenced by Schopenhauer's'.[7] Today, it is rather well known among Wittgenstein's scholars that the Austrian philosopher read Schopenhauer's *The World as Will and Representation* and Schopenhauer did have impact on Wittgenstein, however the scope of this influence is still under discussion.[8] On the other hand, Diamond openly admits that her ethical essays reflect what she has learned from the Austrian philosopher. It may now seem that the discussion of Schopenhauer and Diamond is legitimized *via* the mediation of Wittgenstein – a case of an indirect influence of Schopenhauer on Diamond *via* Wittgenstein – but this relation is somehow more problematic. What philosophers who acknowledge Schopenhauer's impact on Wittgenstein usually have in mind are Wittgenstein's remarks on ethics and aesthetics from the *Tractatus logico-philosophicus*. Meanwhile, Diamond openly rejects this interpretation in favour of James Conant's claim that it is wrong to perceive Wittgenstein's ethical views as influenced by Schopenhauer. Schopenhauer's impact on Wittgenstein is not the main issue of my paper, therefore I cannot discuss it at length and give any conclusive arguments, nonetheless I would like to give one evidence for Diamond's and Conant's view. In *An Introduction to Wittgenstein's Tractatus* G. E. M. Anscombe

---

7   Patrick Gardiner, *Schopenhauer* (London: Penguin Books, 1963), p. 7.
8   For the discussion of Schopenhauer's influence on Wittgenstein see: David Avraham Weiner, *Genius and Talent. Schopenhauer's Influence on Wittgenstein's Early Philosophy* (London and Toronto: Associated University Press, 1992).

wrote: 'As a boy of sixteen Wittgenstein had read Schopenhauer and had been greatly impressed by Schopenhauer's theory of the 'world as idea' (though not of the 'world as will'); 'Schopenhauer' then struck him as fundamentally right, if only a few adjustments and clarifications were made'.[9] There is a strong reason to believe that Anscombe, who was Wittgenstein's pupil and close friend, had first-hand information. If so, the part of Schopenhauer's theory, which Wittgenstein had *not* been greatly impressed by was precisely the part where Schopenhauer discussed his views on aesthetics and ethics – 'world as will'.[10]

It may now seem that by accepting Conant's and Diamond's claim I have just undermined my argument. However, the resemblances between Schopenhauer and Diamond that I am going to discuss in the next part of my paper are not intended to point at some kind of indirect influence of Schopenhauer on Diamond's moral views. Neither am I going to suggest that Diamond's interest in animal ethics is in any way dependent on the fact that Schopenhauer was the first philosopher who created an ethical system acknowledging animals as moral beneficiaries. The resemblances – or rather some common features of corresponding descriptions of the moral relations between people and animals – are rooted in a fact that Wittgenstein's, and therefore also Diamond's, philosophy operate within an anti-intellectualist tradition[11] founded

---

9   Gertrude Elizabeth Margaret Anscombe, *An Introduction to Wittgenstein's Tractatus*, (New York: Harper & Row, Publishers, 1959; reper. 1963), p. 11-12.

10   There is a further problem in sustaining the claim that Wittgenstein was influenced by Schopenhauer's moral views. Namely, it immediately raises a question: which moral views? It is possible to discriminate three stages of Schopenhauer's ethics. The first one, lowermost – which can be found in the *Aphorisms on the Wisdom of Life* – is not ethics in a strict sense, but rather wise tips on how to live a happy life. It is an eudaimonology focused on cultivation of one's own egoism. The second stage is morality in a 'narrower sense'. Here belongs both of Schopenhauer's prize essays *On the Freedom of the Will* and *On the Basis of Morals* which were published together in 1841 under the title of *Die beiden Grundprobleme der Ethik, behandelt in zwei akademischen Preisschriften* (published in English as *The Two Fundamental Problems of Ethics)*. Schopenhauer discussed issues concerning animals in the latter, therefore while discussing Schopenhauer's morality of compassion in the subsequent sections I will refer mostly to this essay. The third stage of morality is morality seen from 'a higher standpoint' and leads to the denial of the Will-to-Live. It can be viewed as a kind of a-theological salvation.

11   The label is Hans-Johann Glock's. The main features of Schopenhauer's anti-intellectualism are instrumental character of the intellect, its secondary position in relation to will, and recognition of thought as a biological function, cf. Hans-

by Schopenhauer. Wittgenstein adopted and improved the descriptive (morphological) parts of Schopenhauer's philosophy,[12] together with borrowing the concept of family resemblance (*Familienähnlichkeit*), but rejected the parts of Schopenhauer's thought, where the German philosopher discussed 'the world as will' – i.e. his moral and aesthetic views. Therefore, although the axiological questions in Schopenhauer and Wittgenstein (and philosophers strongly inspired by Wittgenstein like Diamond) are being asked from a similar, common ground – ground prepared by Schopenhauer – the answers may vary. Since in the following parts I am going to pay attention primarily to the resemblances, I shall put forward briefly one crucial difference as a reminder.[13]

Schopenhauer's dissatisfaction with the morphological method led him to introduce a new philosophical method, sometimes called by scholars *hermeneutics of experience*. Hermeneutics of experience is a very specific kind of understanding constituted by a relation between an intuitive representation (an object) and a subject's will. Schopenhauer expressed it as follows: 'We will be particularly interested in discovering the true *meaning* of intuitive representation; we have only ever *felt* this meaning before, but this has ensured that the images do not pass by us strange and meaningless as they would otherwise necessarily have done; rather, they speak and are immediately *understood* and *have an interest* that engages our entire being'.[14] Wittgenstein, on the contrary, is not concerned with the subject's will as a source of meaning. For him it is rather the sphere of social human practice where what is significant can be observed, recognized and described.

Johann Glock, 'Schopenhauer and Wittgenstein. Language as Representation and Will', *The Cambridge Companion to Schopenhauer*, ed. by Christopher Janaway (Cambridge: Cambridge University Press, 2006), pp. 422-458.

12   For the relation between morphology and aetiology in Schopenhauer's philosophy see: Amadeusz Just, 'Would early Wittgenstein have Understood a Lion?', *Realism – Relativism – Constructivism. Contributions of the 38th International Wittgenstein Symposium*, ed. by Christian Kanzian, Josef Mitterer, Katharina Nages (Kirchberg am Wechsel: Austrian Ludwig Wittgenstein Society, 2015), pp. 148-150.

13   Without a doubt there are more differences that similarities between Schopenhauer's and Wittgenstein's (or Diamond's) philosophies. The aim of this paper is to discuss some resemblances which may point at a common – to a certain extent – shape of thought.

14   Arthur Schopenhauer, *The World as Will and Representation*, vol. 1, trans. and ed. by Judith Norman, Alistair Welchman, Christopher Janaway (Cambridge: Cambridge University Press, 2010), p. 119.

II

At the beginning of this paper I have mentioned two sorts of arguments against cruelty to animals: 1) *indirect claim view* that cruelty to animals may lead to cruelty to human beings, 2) Regan-Singer's *argument from common capacities*. Both of these arguments are constantly in use. The latter is so widespread in everyday life (we all have seen mothers admonishing their children: 'Stop tormenting dog because it hurts him') and public and academic discourse that there is no need to give any further examples. The former is not so common nowadays, however it was explicitly used some years ago in Poland during a public discussion concerning ritual slaughter. Some of the opponents of ritual slaughter claimed that this way of killing animals for meat is so brutal that it lowers the level of human sensitivity and therefore it should be prohibited. Leaving out for a while the actual effectiveness of these arguments, someone could say that any argument, which can persuade at least one person of not killing or hurting animals is worth formulating. Diamond even mentions that it was once argued to her that she 'should not criticize Peter Singer's arguments about vegetarianism because they work'.[15] It looks as if every criticism was considered as detrimental. However, as Diamond noticed, what persuades people is one thing, and what is a good argument is something else: 'We can judge that the proof is sound or that it is confused or fallacious independently of our knowledge of the conditions in which people exposed to it retain and those in which they change their old attitudes'.[16] Therefore, in morality it is necessary to distinguish between what works on people and what is a good argument or a genuinely convincing way of changing someone's way of looking at things. Schopenhauer also emphasised the importance of a good argument as a proper way of changing – or rather activating – moral attitudes,[17] however his perspective slightly differs from Diamond's.

---

15   Diamond, p. 7.
16   Ibid.
17   The possibility of changing someone's moral attitude is yet another issue in Schopenhauer's philosophy. On the one hand Schopenhauer claimed that 'It would be just as absurd to expect our systems of morals and ethics to inspire virtuous, noble and holy men as it would be to think that our aesthetics could create poets, painters and musicians' (*The World as Will and Representation*, p. 298). On the other hand, although our moral character is inborn and unchangeable, the genuine expression and realization of our moral character is also essentially dependent on the motives, arguments and descriptions of

Schopenhauer's aim was to characterise a moral way of acting and to distinguish moral actions from these actions which only superficially seem to be moral. Therefore, actions are *not* moral, which are due to fear of God, legal order, control of public opinion, politeness etc. These are merely different forms of social apparatuses, which aim at restricting human egoism. Schopenhauer claims that 'there is the recognition that a moral way of acting that was set in train merely by threatened punishment and promised reward would be such a thing more in appearance than in reality; because it would surely rest at bottom upon egoism, and then what would be decisive in the final instance would be the greater or lesser ease with which one person rather than the other held beliefs on insufficient grounds.'[18] A deed to be of moral worth cannot be motivated by egoism. Fear of punishment (in this or the other world) or recognition of some advantages (e.g. politeness) can be extremely persuasive and work on people very well, however these motivations are egoistic. By the same token, the *indirect claim view* that cruelty to animals may lead to cruelty to human beings, is a form of an egoistic motivation. It may be understood that 1) cruelty to animals could *make me* a worse person, therefore I should not act this way, or 2) cruelty to animals should be prohibited, because it makes people worse and consequently *makes us* live in a worse society, where *we ourselves* could become victims of cruelty. In both cases the motive is one's own well being. This is the first reason why Schopenhauer rejected the *indirect claim view*. Although the argument that cruelty to animals may lead to cruelty to human beings may be persuasive, neither is it a properly moral argument, nor it genuinely describes the moral relation between people and animals. Before I move to the second reason, I shall go back to Diamond and discuss her response to Regan-Singer's *argument from common capacities*.

III

Diamond summarised the Regan-Singer's argument in one of her best known essays *Eating Meat, Eating People* from 1978 as

---

the world of everyday experience. Therefore, it is possible that even if our moral character is unimpeachable, our deeds are not moral, because the motives of our deeds are amoral or immoral.

18   Arthur Schopenhauer, 'Prize Essay on the Basis of Morals', *The Two Fundamental Problems of Ethics*, trans. and ed. by Christopher Janaway (Cambridge: Cambridge University Press, 2009), p. 119.

follows: 'We must give equal consideration to the interests of any being which is capable of having interests; and the capacity to have interests is essentially dependent only on the capacity for suffering and enjoyment'.[19] In this form of argument human beings are described as 'human animals' on the basis that we evidently share with animals the capacity for suffering and enjoyment. Therefore, what this argument points out, is that by killing animals for food 1) we treat subjects having very similar capacities unequally, 2) we use the word 'animal' to create an artificial difference between humans and animals, which allegedly legitimise our practices. What is wrong in this kind of argumentation is, as Diamond claims, that this 'approach makes it hard to see what is important *either* in our relationship with other human beings *or* in our relationship with animals'.[20] Alice Crary precisely demonstrated that Diamond's main critical targets are Singer's (and Regan's) presuppositions that it is the suffering of other human beings that stops us from killing and eating them.[21] Diamond argues against this general, utilitarian reasoning – we do not kill other people for food, because it makes them suffer, animals are capable of suffering, therefore we ought not to kill animals for food – by putting forward some pictures. We do not eat people who died in accidents. We do not eat our amputated limbs. The fact that we do not eat dead people or amputated limbs is not a consequence of our unwillingness to make people suffer. The idea that someone might be eaten after her death could distress her, but what is distressing here is not the suffering as there would be no suffering at all. On the other hand vegetarians do not eat meat of dead animals even when their death was caused naturally and their meat is of a perfect quality. So, similarly, there are cases where a vegetarian could eat animal's meat without making it suffer, but he would not do it either. What Regan-Singer arguments according to Diamond are missing, are answers to questions like: why a person is not something to eat? why an animal is not something to eat? what is involved in our not eating people? how does it relate to our possible not eating animals?

---

19   Diamond, p. 320.
20   Ibid., p. 321.
21   In her criticism Diamond bases mostly on Singer's statements, notwithstanding her arguments apply also to Regan. For a reconstruction discriminating Singer from Regan see: Alice Crary, 'Humans, Animals, Right and Wrong', *Wittgenstein and the Moral Live: Essays in Honor of Cora Diamond*, ed. by Alice Crary (Cambridge, London: A Bradford Book, The MIT Press, 2007), pp. 381-404.

This is precisely the point containing 'fundamental confusions about moral relations between people and people *and* between people and animals'.[22] In their argumentation Regan and Singer entirely pass over 'fundamental features of our relationship to other human beings which are involved in our not eating them'.[23] Instead, they point at suffering, a capacity common to human and non-human animals, as an alleged significant reason for moral consideration. I have already demonstrated that the possibility of suffering of some living being is not a necessary reason for not eating it. But Diamond's argument is more sophisticated. She does not deny suffering, but argues that the role suffering plays in the argumentation is confused. Some commentators, even if they notice the difference between Schopenhauer and utilitarians, still claim that what is common to both is an orientation towards feelings and sensations of these beings to whom we have moral obligations and, moreover, that this is precisely the reason why Schopenhauer and utilitarians include animals in their moral theories.[24] This view is obviously false as for Schopenhauer – and for Diamond – suffering does not play any significant role in the moral argumentation concerning animals.

IV

The criticism of the *indirect claim view* situates Schopenhauer's animal ethics among *direct claim views*. *Direct claim views* of our moral relations with animals describe animals as direct objects of moral claims. The proponents of the *direct claim view* argue that '[1] our own claim to moral consideration is grounded not in higher capacities of mind like reflection, but rather in lower capacities like sentience (...) [2] consistency obliges us to treat all (human and nonhuman) creatures as imposing equivalent, and equivalently direct, moral claims insofar as they possess the pertinent capacities'.[25] At first glance it seems that Schopenhauer would agree with both aforementioned claims. The German philosopher was the first thinker who created an ethical system acknowledging animals as moral beneficiaries by a re-construction

---

22 Diamond, p. 319.
23 Ibid., p. 322.
24 Cf. Ernst Tugendhat, 'Etyka współczucia; zwierzęta, dzieci, dzieci nienarodzone', *Wykłady o etyce*, trans. by Janusz Sidorek (Warszawa: Oficyna Naukowa, 2004), pp. 185-206.
25 Crary, pp. 382-383.

of the very concept of the subject. To accomplish it, he attacked the philosophy of consciousness, limited the scope of human rationality, which was back then and is often still regarded as a criterion and basis for morality. Further, he embodied the mind and pointed out its dependence on and secondary character to the body understood as embodied will.[26] For the argument it could be said that 1) will (the subject of willing) is a kind of lower capacity akin to sentience, and 2) since will is the essence which is the same for both man and animal – the common capacity – it would be an inconsistency to treat human and animal beings unequally. I have just phrased Schopenhauer's claims in a form of what we today call the *direct claim view*, but the distinction between indirect and direct argumentation is already present in Schopenhauer. As I have already mentioned at the beginning of my paper, Schopenhauer criticized the 'bad argument that cruelty to animals leads to cruelty to human beings' – and this is the second reason why Schopenhauer rejected this argument – because it assumes 'as though man were a direct object of moral duty, the animal being merely indirect, in itself a mere thing'.[27] In the framework of an *indirect claim view*, animals in themselves are not objects of moral concern or moral action. Their status is not different from the status of a thing. It is equally bad to be cruel to animals as to play with sharp tools – both of these actions could lead to hurting a human being. In other words, in the framework of *indirect claim views* there is no place for any kind of moral treating of animals, which could recognize them as objects of moral consideration. Yet, although animals are not different from things, the proponents of these view feel the need to morally distinguish our deeds toward animals and toward things. But to do so, they would have to present some kind of moral consideration of animals, which is not possible on the basis of their ethics. Therefore, arguments of such a form are incorrect, however their incorrectness directs our attention to our genuine moral attitude toward animals. By contrast, in Schopenhauer's moral theory, animals are direct objects of morality. Moreover, in Schopenhauer's ethics, animals are included into the realm of morality, the realm of moral community, they are *inside* ethics. This would be the third distinctive feature of Schopenhauer's

---

26   Schopenhauer turned the relation between the subject of knowing and the subject of willing upside down. We are not capable of subordinating our will to ourselves, our subject of knowing cannot govern our will or impose laws on it, as it is precisely our will, which for the most time rules our cognition. It is important to note that Schopenhauer's use of the term *will* is idiosyncratic.

27   *Parerga and Paralipomena*, p. 372.

moral theory, the feature that situates Schopenhauer not only against the *indirect claim view* but also against the Regan-Singer arguments from common capacities (an example of a *direct claim view*) on the one hand and on the other, closer to Diamond's ethics. However, I cannot directly elaborate in this paper on the distinction between situating animals inside/outside ethics.[28] Instead I will demonstrate why phrasing Schopenhauer in a form of *direct claim view* can only lead to confusions and, thus, what distinguishes his view from utilitarian argumentations.

The aforementioned book – *Die beiden Grundprobleme der Ethik* – which Schopenhauer sent to Perner, contained two essays, one of which was the *Prize Essay on the Basis of Morals*. Roughly speaking, it consists of two parts. The negative one, where Schopenhauer discussed a whole tradition of ethical systems – mostly Kant's categorical imperative. He explicitly differentiates his ethics from ethical systems of former philosophers in that it describes *how* humans in reality *act* and not how humans *ought to* act – so it is descriptive, not prescriptive. This is yet another feature common to Schopenhauer and Diamond. The American philosopher objects to the attempts in our ethical reflection to lay down philosophical requirements. This is expressed by the construction of arguments under the influence of some supposed ideal and 'the requirements which we [by that means] lay down stop us seeing what moral thought is like'.[29] In the second, positive part of the *Prize Essay on the Basis of Morals* Schopenhauer introduces his basis of morality – compassion (*Mitleid*).[30] Compassion is not a feeling. Next to egoism and

---

28 The distinction between situating animals (or human beings) inside/outside ethics is originally Alice Crary's. My intention was only to suggest that Schopenhauer's moral views comply more with what Crary describes as situating animals *inside* ethics. For Crary's inspiring argumentation see: Alice Crary, *Inside Ethics. On the Demands of Moral Thought* (Cambridge, London: Harvard University Press, 2016).

29 Diamond, p. 23.

30 Schopenhauer brings forward his moral theory without any specific reference to animals. Only on few pages of § 19 *Confirmations of the Foundation of Morals Expounded* he demonstrates that morality of compassion applies equally to humans and animals. I have decided to discuss Schopenhauer's ethics only in this regard as it is necessary for my argumentation. For a detailed discussion see: David E. Cartwright, 'Compassion and Solidarity with Sufferers: The Metaphysics of *Mitleid*', *Better Consciousness: Schopenhauer's Philosophy of Value*, ed. by Alex Neill, Christopher Janaway (West Sussex: Wiley-Blackwell, 2009), pp. 138-156; David E. Cartwright, 'Schopenhauer's Narrower Sense of Morality', *The Cambridge Companion to*

malice, compassion is one out of three fundamental *springs* of *human conduct*, whose motive is others' well being. The question is: who can be an object of compassion? In the § 16 *Presentation and Proof of the Sole Genuine Moral Incentive* Schopenhauer sets down 9 premises of the proof. These premises consist of something akin to transcendental discussion of compassion demonstrating its necessary conditions. Two of them refer to the possible object of compassion: '3) What moves the will is solely well-being and woe as such, and taken in the widest sense of the word; just as conversely well-being and woe means 'in accordance with a will, or against it'. Thus every motive must have a relation to well-being and woe. 4) Consequently, every action relates to a being receptive to well-being and woe, as its final end'.[31] As we can read in the third premise, good and bad are defined in relation to a willing subject. Good is something which is in conformity with a will and bad is something that is contrary to it. Since compassion is motivated by someone else's well being or woe, a possible object of compassion must be a being sentient to well-being and woe. The 'receptiveness to well-being and woe' makes Schopenhauer a representative of *direct claim view*. However, contrary to Regan and Singer, sentience to well-being and woe is *not* a capacity common to humans and animals *that* makes them object of moral consideration. It is rather a necessary feature owing to which it makes sense to talk about compassion, which makes the use of the notion *compassion* meaningful (it becomes clearer, when we read Schopenhauer's remark as a grammatical rule in Wittgenstein's sense).[32] However Schopenhauer's transcendental perspective discusses the *possibility* of compassion, not a *use* of the word *compassion*. No matter how we intend to read Schopenhauer's premises, sentience to well-being and woe (feelings of pleasure and pain) is not a reason to treat some being in such-and-such a way. In a previous section I have mentioned Diamond's claim that suffering does not play any significant role in the moral argumentation concerning animals, however to 'doubt about that is, in most ordinary cases, as much out of place as it is in

---

    Schopenhauer, ed. by Christopher Janaway (Cambridge: Cambridge University Press, 2006), pp. 252-292.

31   'Prize Essay on the Basis of Morals', p. 198.

32   Cf. Wittgenstein's remarks on the use of word *pain*, Ludwig Wittgenstein, *Philosophical Investigations*, trans. by Gertrude Elizabeth Margaret Anscombe, Peter Hacker and Johann Schulte (Malden: Wiley-Blackwell, 2009), pp. 281-288.

many cases in connection with human beings.[33] It would be possible to formulate several arguments in harmony with Diamond's philosophical views to elucidate this issue. I shall turn to only one of them common to both Diamond and Schopenhauer, which is connected to a preoccupation with the method of science in philosophy.

Neither Diamond, nor Schopenhauer opposed science. What they were against was using scientific method in philosophy, and particularly in ethics. No scientific discovery can conclusively inform us on a proper ethical attitude towards animals as ethics is not something scientific. Our moral concepts and the concepts by which we describe our moral lives or actions are not directed towards genetic or neurophysiologic facts like, for instance, scientifically defined pain, suffering or enjoyment. These facts are not significant to us as moral beings. This is precisely the reason why Schopenhauer criticized Darwin's theory for 'being plain empiricism, insufficient in this matter'.[34] The world, apart from its physical meaning discovered by science, also has a moral meaning. But the moral meaning of the world cannot be recognized by the materialistic systems since 'with materialistic systems that represent the world as arising out of a matter endowed with merely mechanical properties and in accordance with the laws of such matter, neither the universal and marvellous appropriateness of nature, nor the existence of knowledge wherein even that matter is first exhibited, is in agreement'.[35] We want to believe that one day science will provide us with all the answers. But the answers which we are looking for in ethics – for instance, how should we treat animals? or what capacity grounds moral considerations? – are out of its range. As Diamond noticed, 'we fail to distinguish between 'the difference between animals and people' and 'the differences between animals and people'.[36] The scientific evidence is not going to show us that 'the difference' between us and animals is not as big as we thought. It can only show, that 'the differences' between us and animals are less sharp than we think. We

---

33 Diamond, p. 320.
34 Several months before his death Schopenhauer read in *Times* a detailed summary of Charles Darwin's *On the Origin of Species by Means of Natural Selection* and commented on it in a letter to Adam von Doβ from 1st March 1860. Quotation after: Ortrun Schulz, *Schopenhauer's Biophilosophy* (Norderstedt: BoD, 2014), p. 9.
35 Arthur Schopenhauer, *Parerga and Paralipomena*, vol. 1, trans. by Eric F. J. Payne (Oxford: Clarendon Press, 1974; repr. 2000), p. 67.
36 Diamond, p. 324.

learn 'the difference' between us and animals in practice, and it is being expressed by our practices. It is 'a central concept for human life and is more an object of contemplation than observation'.[37]

V

In the previous section I have demonstrated why it can be confusing to phrase Schopenhauer's remarks on animal ethics in the form of a *direct claim view*. Schopenhauer did not argue that being sentient to well-being and woe is a capacity common to human and nonhuman animals, therefore we should treat them equally. Nor did he argue – like Singer – that suffering grounds moral considerations. I would suggest that he would have agreed with Diamond on her allegedly controversial claim that 'We cannot point and say 'This thing (whatever concepts it may fall under) is at any rate capable of suffering, so we ought not to make it suffer.'[38] What is wrong with thinking about the relations between human beings and animals through that claim becomes more apparent in another formulation. 'Well, here we have me the moral agent and there we have it, the thing capable of suffering' and pulling out of that 'well, then, so far as possible I ought to prevent its suffering.'[39] The discrimination brought forth by Diamond on the moral agent and a thing possessing pertinent capacity and therefore – through that capacity – becoming an object of moral consideration is rather concealed in Singer-Regan's reasoning. The obvious asymmetry between the two subjects of moral relation leads to questions concerning the actual function of the concept of equality in this view and the relation between human beings and animals that stands behind it. The second problem with this reasoning concerns the phrase 'therefore ought to'. Diamond makes it clear in yet another essay, *The difficulty of reality and the difficulty of philosophy*,[40] where she discussed '*therefore*-arguments'. *Therefore*-arguments begin with a description of a subject and go to its 'therefore', which depends on the former description, description, which is arbitrary and can be easily opposed. In this case, for instance: Animals don't

---

37   Ibid.
38   Ibid., p. 325.
39   Ibid., p. 333.
40   Cora Diamond, 'The Difficulty of Reality and the Difficulty of Philosophy', *Philosophy and Animal Life*, Stanley Cavell... [et al.] (New York: Columbia University Press, 2008), pp. 43-89.

have consciousness, therefore we can do with them whatever we like. Therefore-arguments contain yet another element, namely 'therefore we ought or ought not to'. For Wittgenstein an answer to any ethical law of the form 'Thou shalt...' is: 'And what if I do not do it?'[41] Similarly Schopenhauer criticized any ethics of such a form. He claimed that it can be used only if we presuppose the existence of some external, alien will of which human being is dependent, and which is giving prizes and setting penalties to him. He meant all theological ethics whose foundations lie in the concept of God. The third issue concerns the generality of such an argument, that it is allegedly about every animal and is addressed to all humans. But, as Diamond wrote: 'Animals are not given for our thought independently of such a mass of ways of thinking about and responding to them'.[42]

VI

For the last part of my paper I shall turn to some positive points of Diamond's and Schopenhauer's argumentation. Schopenhauer's descriptive ethics does not prescribe compassion towards animals, but it *recognizes* and *acknowledges* that we are compassionate towards animals. It is a recognition that we *are* compassionate toward animals, therefore – but this is a different use of 'therefore' than in '*therefore*-arguments' – we *can* be compassionate toward animals. Similarly Diamond, by introducing a term 'fellow creature', intends to recognize and acknowledge features of our moral relationship with animals. The process of recognition is related to the fact mentioned in the previous quotation that animals are given for our thought in a variety of ways of thinking about them and responding to them. Schopenhauer was

---

41   Ludwig Wittgenstein, *Tractatus logico-philosophicus*, trans. by David F. Pears and Brian F. McGuinness (London and New York: Routledge, 1961; repr. 2002), 6.422. Another interesting background reading of an author important for Diamond would be Anscombe's essay *Modern Moral Philosophy*, where Anscombe claimed that 'the concepts of obligation, and duty – *moral* obligation and *moral* duty, that is to say – and of what is *morally* right and wrong, and of the *moral* sense of 'ought', ought to be jettisoned' (Gertrude Elizabeth Margaret Anscombe, 'Modern Moral Philosophy', *Human Life, Action and Ethics. Essays by G. E. M. Anscombe*, ed. by Mary Geach and Luke Gormally (Exeter: Imprint Academic, 2005), p. 169).

42   Diamond, p. 331.

probably the first one who recognized the importance of language as a mechanism of creating the *difference* between human and non-human animals. He noticed that when it comes to the natural phenomena like eating, pregnancy, death etc., our language differs depending on whether we talk about human beings or animals. Diamond makes a similar point when she describes different ways in which we relate to animals, as for instance our different attitude towards pets and vermin. By noticing that we do not eat our pets Diamond means that it is not our practice with animals that we call pets, to eat them (if someone claims that he eats his pets, he is making a crude joke).

The notion of a fellow creature or of a living creature that Diamond introduced should not be understood as a biological fact that animals are creatures that are alive. Diamond – and this refers also to Schopenhauer – wants us to look at animals differently, to look at them as at our companions and to respond to them as to our fellows in morality. Schopenhauer gave many examples of situations in which animal beings were looked at as our companions in morality. One of the most suggestive is a story described by William Harris who 'in 1836 and 1837 travelled deep into the interior of Africa solely to enjoy the pleasures of the hunt. In the book of his travels that appeared in Bombay in 1838 he recounts that after he had bagged the first elephant, which was a female, and sought out the fallen animal the following morning, all other elephants had fled the area: only the fallen elephant's young one had spent the night with its dead mother, and now, forgetting all fear, it came towards the hunters giving the liveliest and clearest testimony of its inconsolable misery, and embraced them with its little trunk so as to call on their help. Then, says Harris, true remorse for his deed seized him and it felt to him as if he had committed a murder'.[43] The story is a good example of Diamond's insight, that to think about animals by using such concepts as *charity*, *justice*, *independent life*, *friendship*, *respect*, *pity*, *sparing someone's life*, instead of rational argumentation, may help us not to distance ourselves from animals and the bodily life we share with them.

For both Schopenhauer and Diamond morality is our attitude toward world and life, which is being expressed by our actions and our words. Without a doubt we could change someone's behaviour by establishing a law with severe penalties, but it would not be a change in his or her morality. By giving examples, pictures and stories the two philosophers

---

43   'Prize Essay on the Basis of Morals', p. 229.

want us to change our way of looking at animals, to recognize and acknowledge the genuine moral relations between human beings and animals instead of laying down philosophical requirements of how they ought to be.

Bibliography

Anscombe, Gertrude Elizabeth Margaret, *An Introduction to Wittgenstein's Tractatus*, (New York: Harper & Row, Publishers, 1959; repr. 1963)

Anscombe, Gertrude Elizabeth Margaret, 'Modern Moral Philosophy', *Human Life, Action and Ethics. Essays by G. E. M. Anscombe*, ed. by Mary Geach and Luke Gormally (Exeter: Imprint Academic, 2005), pp. 169-194

Broadie, Alexander, and Pybus, Elizabeth M., 'Kant's Treatment of Animals', *Philosophy*, vol. 49, no. 190 (Oct., 1974), pp. 375-383.

Cartwright, David E., 'Compassion and Solidarity with Sufferers: The Metaphysics of *Mitleid*', *Better Consciousness: Schopenhauer's Philosophy of Value*, ed. by Alex Neill, Christopher Janaway (West Sussex: Wiley-Blackwell, 2009), pp. 138-156

Cartwright, David E., 'Schopenhauer's Narrower Sense of Morality', *The Cambridge Companion to Schopenhauer*, ed. by Christopher Janaway (Cambridge: Cambridge University Press, 2006), pp. 252-292

Crary, Alice, 'Humans, Animals, Right and Wrong', *Wittgenstein and the Moral Live: Essays in Honor of Cora Diamond*, ed. by Alice Crary (Cambridge, London: A Bradford Book, The MIT Press, 2007), pp. 381-404

Crary, Alice, *Inside Ethics. On the Demands of Moral Thought* (Cambridge, London: Harvard University Press, 2016)

Diamond, Cora, 'The Difficulty of Reality and the Difficulty of Philosophy', *Philosophy and Animal Life*, Stanley Cavell... [et al.] (New York: Columbia University Press, 2008), pp. 43-89

Diamond, Cora, *The Realistic Spirit. Wittgenstein, Philosophy, and the Mind* (Cambridge, London: A Bradford Book, The MIT Press, 1991)

Gardiner, Patrick, *Schopenhauer* (London: Penguin Books, 1963)

Glock, Hans-Johann, 'Schopenhauer and Wittgenstein. Language as Representation and Will', *The Cambridge Companion to Schopenhauer*, ed. by Christopher Janaway (Cambridge: Cambridge University Press, 2006), pp. 422-458

Just, Amadeusz, 'Would early Wittgenstein have Understood a Lion?', *Realism – Relativism – Constructivism. Contributions of the 38th International Wittgenstein Symposium*, ed. by Christian Kanzian, Josef Mitterer, Katharina Nages (Kirchberg am Wechsel: Austrian Ludwig Wittgenstein Society, 2015), pp. 148-150

Libell, Monica, *Morality Beyond Humanity. Schopenhauer, Grysanowski, and Schweitzer on Animal Ethics* (LAP: Saarbrücken, 2010)

Magee, Bryan, *The Philosophy of Schopenhauer* (Oxford, New York: Clarendon Press, Oxford University Press, 1983; repr. 2009)

Schopenhauer, Arthur, *Parerga and Paralipomena*, vol. 1, trans. by Eric F. J. Payne (Oxford: Clarendon Press, 1974; repr. 2000)

Schopenhauer, Arthur, *Parerga and Paralipomena*, vol. 2, trans. by Eric F. J. Payne (Oxford: Clarendon Press, 1974; repr. 2000)

Schopenhauer, Arthur, 'Prize Essay on the Basis of Morals', *The Two Fundamental Problems of Ethics*, trans. and ed. by Christopher Janaway (Cambridge: Cambridge University Press, 2009), pp. 113-258

Schopenhauer, Arthur, *The World as Will and Representation*, vol. 1, trans. and ed. by Judith Norman, Alistair Welchman, Christopher Janaway (Cambridge: Cambridge University Press, 2010)

Schulz, Ortrun, *Schopenhauer's Biophilosophy* (Norderstedt: BoD, 2014)

Tugendhat, Ernst, 'Etyka współczucia; zwierzęta, dzieci, dzieci nienarodzone', *Wykłady o etyce*, trans. by Janusz Sidorek (Warszawa: Oficyna Naukowa, 2004), pp. 185-206

Weiner, David Avraham, *Genius and Talent. Schopenhauer's Influence on Wittgenstein's Early Philosophy* (London and Toronto: Associated University Press, 1992)

Wittgenstein, Ludwig, *Philosophical Investigations*, trans. by Gertrude Elizabeth Margaret Anscombe, Peter Hacker and Joachim Schulte (Malden: Wiley-Blackwell, 2009)

Wittgenstein, Ludwig, *Tractatus logico-philosophicus*, trans. by David F. Pears and Brian F. McGuinness (London and New York: Routledge, 1961; repr. 2002)

BENEDETTA PIAZZESI
SCIENTIFIC BESTIARIUM:
THE LIVING, THE DEAD, AND THE NORMAL

My aim is to analyse scientific literature and its representation of the animal body in relation to the disciplinary institutions of its time, namely zootechnics. I will focus on the nineteenth century as the moment of birth of a specific biological discourse and as the moment of deployment of the Industrial Revolution, which had a significant impact on animal breeding. This conjuncture produces a radical new image of the animal body and of animality in general, which plays an important role not only in science and zootechnics, but also in philosophy and the human sciences. I will frame the evolution of scientific discourses on the animal body from the Greeks to the Modern Age, in order to present their material history in relation to the concrete practices that involved animals in their time. I will finally focus on two of the most important scientific models of the nineteenth century – Pasteurism and Darwinism – as cutting-edge moments in the history of biology, precisely due to their innovative relation to the zootechnical institution and its related conceptualization of the body.

Material Conditions of Scientific Discourse: Which Bodies? Which Animals?

> Nothing is as important for a biologist as his choice of material to study
>
> – Georges Canguilhem

The conditions of possibility which gave rise to the new scientific branch of biology between the eighteenth and nineteenth centuries have been retraced by epistemologists and historians of ideas such as Georges Canguilhem and Michel Foucault. They tracked these conditions in their multiplicity, especially in their interaction with other scientific disciplines and human sciences. I will use this general frame, but I will focus on the particular relation of biology with one specific element of its genealogy, namely the zootechnical institution. Zootechnics has been unexpectedly omitted in the epistemological and genealogical project of critical theories but, in an attempt at providing a materialistic genealogy of knowledge, it should clearly be one of the most important places of observation and transformation of living beings in modern society.[1] If the animal body is the material upon which biology works, its subject matter of research, then one should ask: how did its transformation within breeding farms enable a new form of knowledge?

The importance of the problem is noted by Georges Canguilhem himself when he says: 'We know that nothing is as important for a biologist as his choice of material to study'.[2]

What kind of animals have been observed and examined by ancient naturalists and then by the first modern scientists? What kind of bodies were produced by industrial breeding, and provided by nineteenth-century biologists? And finally, what kind of body-models are required by contemporary laboratories that reproduce and raise animals on their own?

Ancient Animals from Techne to Aletheia

> Una disponibilità ad uccidere senz'altro scopo che la conoscenza[3]
>
> – Mario Vegetti

Scholars of antiquities recognise the cutting-edge role of Aristotle in the history of zoological knowledge. In particular, Mario Vegetti[4] identifies two

---

1   Anita Guerrini points out a similar lack of attention to animals' role in the history of society, and in the history of sciences in particular: 'While the development of the field of 'animal studies' has certainly helped to insert animals into scholarly discourse, I believe it has also defined the field in a way that separates it from other disciplines, so that ignoring animals as historical actors remains the norm outside of that small subdiscipline', Anita Guerrini, *The Courtiers' Anatomists: Animals and Humans in Louis XIV's Paris* (Chicago: The University of Chicago Press, 2015), p. 3.

2   Georges Canguilhem, *Knowledge of Life*, trans. by Stefanos Geroulanos and Daniela Ginsburg (New York: Fordham University Press, 2008), p. 11.

3   'A willingness to kill with no other purpose apart from knowledge', M. Vegetti, *Il coltello e lo stilo*, cit., p. 41 [My translation].

4   Mario Vegetti, *Il coltello e lo stilo. Le origini della scienza occidentale* (Milano: il Saggiatore, 1996).

different stages in the Stagirite's biological production. In the first stage, which corresponds to the *Historia animalium*, Aristotle uses 'second hand' information about animals, which the so-called *technitai* – technicians of animal handling (namely hunters, fishers, breeders, butchers, etc.) – collected over years. Aristotle's work consisted in organizing an incredible amount of varied information, mainly concerning the phenomenology of living animals, and transposing them into the ennobling space of the written composition. This is a foundational gesture for science, because it combines empirical and theoretical tools. Aristotle frequently states the necessity of an empirical approach to the natural sciences, but his *empeiria* is far from that of the modern experimental method. *Empeiria* is rather the practical expertise that an artisan acquires through familiarity with a matter and work. This practical origin, arising from technicians' hands, ensures that Aristotle's first classification – and the drafts of zoological classification typical of the pre-Aristotelian philosophy – were organized under morphological, ethological and ecological criteria.[5] These elements of information fundamentally concern the moment of the capture or the breeding operations, and consequently produce a description and taxonomy of living.

Nineteenth-century French epistemology tries to rewrite the history of science precisely in its pragmatic origins. Georges Canguilhem states that it is not science that produces technology, but that we should forge a genealogy of theoretical knowledges from their material conditions of possibility, and show how 'la biologie est peu à peu sortie des pratiques de chasse, d'élevage, de médecine, d'agriculture'.[6] This utilitarian relationship with reality and the living is typical of the technical gesture, and it cannot but orient the organization of knowledge. Excising this pragmatic origin and this utilitarian leaning is one of the principle operations of science, in its universalistic and objective aspirations.

In the *Historia animalium*, instead, Aristotle allows us to easily penetrate the teleological and anthropological orientation of taxonomy. What is remarkable is that, in these very early researches, he already moves towards something different: he declares the necessity, in some

---

5   Pierre Pellegrin, *La classification des animaux chez Aristote* (Paris: Les Belles Lettres, 1982); Arnaud Zucker, *Aristote et les classifications zoologiques* (Paris: Peeters, 2005).

6   '[B]iology gradually arises from hunting, farming, medicine and agriculture practices', Georges Canguilhem, *Activité Technique et Création*, in *Œuvres complètes, Tome I, Écrits philosophiques et politiques (1926-1939)* (Paris: Vrin, 2011), p. 504 [My translation].

cases, of resorting to an autoptic perspective on animals' internal composition.[7] The scholars should observe corpses not just as fishers and butchers do, but rather considering them as animals killed following a procedure which has the sole purpose of knowledge.

> Now, as the nature of blood and the nature of the veins have all the appearance of being primitive, we must discuss their properties first of all, and all the more as some previous writers have treated them very unsatisfactorily. And the cause of the ignorance thus manifested is the extreme difficulty experienced in the way of observation. For in the dead bodies of animals the nature of the chief veins is undiscoverable, owing to the fact that they collapse at once when the blood leaves them; for the blood pours out of them in a stream, like liquid out of a vessel, since there is no blood separately situated by itself, except a little in the heart, but it is all lodged in the veins. In living animals it is impossible to inspect these parts, for of their very nature they are situated inside the body and out of sight. For this reason anatomists who have carried on their investigations on dead bodies in the dissecting room have failed to discover the chief roots of the veins, while those who have narrowly inspected bodies of living men reduced to extreme attenuation have arrived at conclusions regarding the origin of the veins from the manifestations visible externally. [...] The investigation of such a subject, as has been remarked, is one fraught with difficulties; but, if any one be keenly interested in the matter, his best plan will be to allow his animals to starve to emaciation, then to strangle them on a sudden, and thereupon to prosecute his investigations.[8]

These kinds of methodological statements are very important to understand how Aristotle unifies the technical culture with the theoretical and sacred one of the Pythagoreans. He inherits the idea of a disinterested knowledge from the Pythagoreans, but adds familiarity with the animals' bodies and their willingness to kill. As a result, he will develop this idea of killing for pure theory.[9]

Actually Aristotle includes dissection within the Greek theoretical concept of truth as disclosure, *aletheia*. Through a practical penetration,

---

7   It seems that Alcmaeon of Croton was the first physician to practice anatomical dissection in the fifth century BCE.

8   Aristotle, *History of Animals*, trans. by D'Arcy Wentworth Thompson (Oxford: Clarendon Press, 1910), III 2-3, 511b13-513a12.

9   Vegetti, p. 41.

anatomy reveals the original manifestness of being. On several occasions Aristotle professes that reality shows itself a truth that the scientist has only to register. Later we will reconnect with this Greek concept of *aletheia* and its claim to facticity that characterizes the scientific discourse up to our time.

In the remaining biological treatises – *De Partibus animalium* and *De Generatione animalium* among others – Aristotle starts to implement this innovative epistemological effort. He refuses to mention his technical sources, and declares the necessity of a re-foundation of knowledge on a more 'theoretical' method: from non-utilitarian dissection, to mere cognitive purposes. As Maurice Manquat says, Aristotle starts 'classifying corpses', which is a gesture of great importance, because it points to anatomy as the preferred taxonomic parameter, which will be definitively undertaken in the Modern Age. The simple observation of living animals produced a complex description and organization of ancient animals as 'behavioural morpho-types', at the intersection of morphological, ethological, ecological and symbolic characters,[10] whereas the complex artificial gesture of dissection allows a more coherent and plain description of animals through their hidden essential bodily structure. This inner structural description is much more effective for a comparative organization of species, a fact that places Aristotle among the moderns, for it seems to presage the possibility of a universal taxonomy. Especially in *De Partibus* and *De Generatione* Aristotle seems to be interested not in single animals, but rather in the 'general animal'[11] which can be outlined through comparative anatomy.

---

10   Cf. Pietro Li Causi, 'Corpi, spazi, luoghi, animali. La zoologia dei greci dall'animale come spazio visivo localizzato alle funzioni dell'anima', in *Athenaeum*, 96, 1, 2008, pp. 55-75.

11   '[C]'est l'animal en général dont il fait l'histoire', Armand-Gaston Camus (ed. by), *Histoire des Animaux d'Aristote* (Parigi, 1783), p. 13. The 'general animal' had been a stand-in for human anatomical models for long time, as Guerrini states: 'Animals had been accepted as stand-ins for humans in anatomical study since at least the time of Aristotle. Many acknowledged that animals were not perfect models for humans; Vesalius had criticized Galen on precisely this point, and indeed Galen knew quite well that animals were not always ideal. But there was widespread acceptance of the general principle that animal bodies – some more than others – were similar enough to humans to act as stand- ins', Guerrini, p. 4.

The Enlightenment of Corpses

> Open some cadavers, and you will soon see disappearing
> the obscurity which observation alone could not have dispelled
> – Xavier Bichat

From the first animal dissections by Alcmaeon of Croton in the fifth century BCE and the Aristotelian dissective *aletheia*, anatomy started to emerge as a great exploratory method. Aristotle produced a set of lost illustrations which he frequently refers to as *Anatomai*. These would have been invaluable to understanding the visual consciousness of his time and its contribution to the Greek idea of *theoria*, as etymologically connected to the operation of seeing.[12] The importance of the Greek heritage of eyesight is preserved and increased in the modern conception of truth, culminating in the age of *Aufklärung*, which maintains this idea in its own name: *enlightenment*.[13]

Anatomy has an important role in this theoretical and moral process of dissipation of darkness. The seventeenth-century anatomist Jean Riolan le Jeune defines dissection the 'seeing hand' of medicine.[14] From the Renaissance onward, and in seventeenth century more than ever, it 'has a justifiable claim to be the most widespread and significant scientific activity'.[15]

Anatomy finds in graphical illustrations its textual theoretical counterpart and support, as it is possible to consider through the well-known work of the Italian anatomist Carlo Ruini who, in 1598, drafted very detailed anatomical tables, completely dedicated to equine species (image #1).[16]

---

12   Cf. Andrea L. Carbone, *Aristote illustré. Représentatio du corps e schématisation dans la biologie aristotélicienne* (Paris: Classiques Garnier, 2011).
13   'The eye becomes the depositary and source of clarity; it has the power to bring a truth to light that it receives only to the extent that it has brought it to light; as it opens, the eye first opens the truth: a flexion that marks the transition from the world of classical clarity – from the 'enlightenment' – to the nineteenth century', Michel Foucault, *The Birth of the Clinic*, trans. by Alan Sheridan (London: Routledge, 2003), p. xiii.
14   Cited in Guerrini, p. 38.
15   Guerrini, p. 5. René Descartes himself emphasises the importance of anatomy among scientific practices, in several lettres, especially addressed to Mersenne [18 december 1629; 15 april 1630; 29 january 1639; 20 february 1639, etc...].
16   Carlo Ruini, *Anatomia del cavallo infermità et suoi rimedii* (Venice: Fioravante Prati, 1618).

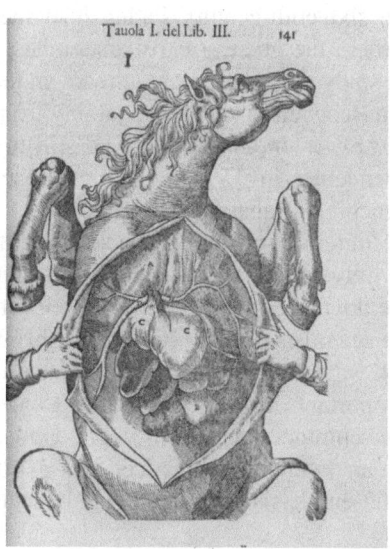

By contrast, the history of human anatomy remained far more troublesome until its affirmation in the eighteenth century, due to religious and more general cultural interdictions against handling and opening human corpses.[17] There were, of course, some important ground-breaking exceptions: the *Anathomia Corporis Humani*,[18] written and illustrated in 1316 by Mondino de' Liuzzi, founder of the first European School of Human Anatomy at the University of Bologna; Leonardo da Vinci's brilliant anatomical draft where he tried to exceed the classical pictorial interest for surface anatomy of his masters;[19] and finally, the sixteenth-century Flemish anatomist Andreas Vesalius, who is universally considered to be the founder of modern anatomy.[20]

---

17   Andrew Cunningham, *The Anatomical Renaissance: The Resurrection of the Anatomical Projects of the Ancients* (Aldershot: Scolar Press, 1997). Roger French, *Dissection and Vivisection in the European Renaissance* (Aldershot: Ashgate, 1999).

18   Mondino de' Liuzzi, *Anothomia*, trans. by Piero Giorgi and Gian Franco Pasini (Bologna: Istituto per la storia dell'Università di Bologna, 1992).

19   Giuseppe Olmi, *Representing the Body: Art and Anatomy from Leonardo to Enlightenment* (Bologna: Bononia University Press, 2006).

20   Andreas Vesalius, *On the Fabric of the Human Body*, trans. by William Frank Richardson and John Burd Carman, 5 vols, (San Francisco and Novato: Norman Publishing, 1998-2009). Marion Harry Spielman, *The Iconography*

Throughout the sixteenth-century Padua held the most important medical school, where the greatest early anatomists, such as Vesalius and Bauhin, could study and teach. Here, already in the early sixteenth-century Alessandro Benedetti had taught anatomy from the tables of the first temporary *Theatrum Anatomicum*.[21] And still in Padua Girolamo Fabrici d'Acquapendente, in 1595, constructed the first permanent structure for the teaching of anatomy.

But anatomy definitely flourished in the seventeenth and eighteenth centuries. Even if Italy cannot claim a central position in a geography of modern science anymore, due to the more relevant experiences of British and French scientists,[22] it still had part in this process.

The well-known wax anatomical models realized in the eighteenth-century within important institutions thanks to the partnership of surgeons and wax sculptors such as Giuseppe Galletti and Clemente Susini[23] represent an interesting practice, and a prosecution of the visual and aesthetic emphasis typical of the sixteenth-century Italian anatomists.

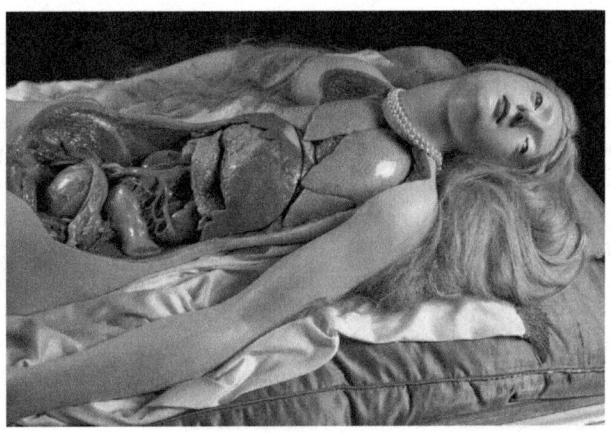

of *Andreas Vesalius (André Vésale), Anatomist and Physician, 1514-1564* (London: John Bale, Sons & Danielsson, 1925).

21   Alessandro Benedetti, *Historia corporis humani sive Anatomice* (Venezia: Bernaldino Guerraldo Vercellensi, 1502).

22   For an accurate survey of French anatomical studies in the seventeenth century, cf. the aforementioned Guerrini.

23   Georges Didi-Huberman, *Ouvrir Vénus. Nudité, rêve, cruauté* (Paris: Gallimard, Parution, 1999).

In *The Birth of the Clinic* Michel Foucault focuses on this crucial passage from eighteenth-century Italian anatomists such as Antonio Maria Valsava and Giovanni Battista Morgagni to French Enlightenment biologists, and the raising of a new medical framework built around pathological anatomy. The surgeon and biologist Marie François Xavier Bichat and his work *Anatomie Générale* (1801) best represent this medical transformation, the role of dissection, and its philosophical implications. Foucault defines the anatomical metaphysics that takes shape as 'mortalism': that particular kind of vitalism which defined life by its contrary, namely as 'the sum of all those forces which resist death'.[24]

The problem of life and that of pathology will be read through the prism of death:

> It is from the height of death that one can see and analyse organic dependences and pathological sequences. [...] The living night is dissipated in the brightness of death [...]. [I]n the boldness of the gesture that violated only to reveal, to bring to the light of day, the corpse became the brightest moment in the figures of truth. [...] That which hides and envelops, the curtain of night over truth, is, paradoxically, life; and death, in the contrary, opens up to the light of day the black coffer of the body: obscure life, limpid death, the oldest imaginary values of the Western world are crossed here in a strange misconstruction that is the very meaning of pathological anatomy.[25]

It seems that biology has to look at the varied phenomenology of vital processes through the motionless figure of the corpse, to adopt the peculiar rigour of science, which is literally a *rigor mortis* in this case.

---

24   Marie François Xavier Bichat, *General Anatomy Applied to Physiology and Medicine*, trans. by Constant Coffyn (London: King's College, 1824), p. 1.

25   Foucault, pp. 145-166.

Wild Tests and Biting Death

> Quanto si trovino belle le viscere degli animali fatti morir di fame[26]
>
> – Francesco Redi

The need for stillness and constancy, manifested in the research for death of the anatomical glance, should be placed in a broader frame, which is that of an epistemology of modern sciences marked by the affirmation of the experimental method. Initially developed among physical and chemical sciences, it is finally acquired by biology over a long process that covered XVII, XVIII and XIX century. I will focus on two of the major scholars and pioneers of experimental biology, Francesco Redi and Louis Pasteur, who, at a distance of 200 years, laid the foundations for parasitology and epidemiology, disciplines that have been of great importance for the management of animal and human populations, that is to say, for biopolitics.[27] I will consider some epistemological transformations through these centuries, by focusing on the description of their experiments on animals.

In a text of 1684, *Osservazioni intorno agli animali viventi che si trovano negli animali viventi*,[28] Francesco Redi records the descriptions and the illustrations of more than 100 parasites. The treatise gives an overview of the various genres of worms that lived in the different species of animals he observed and dissected: reptiles, hedgehogs, martens, lions, deers, dogs, humans, cats, rabbits, but also dolphins, fishes, and more. The variety of his *bestiarium* is astonishing.[29] For most of these animals he illustrates the results of several dissections, aimed to discover their inner parts and their animal inhabitants.

---

26   'How beautiful are the intestines of the animals starved to death', Francesco Redi, *Osservazioni intorno agli animali viventi*, in *Opere di Francesco Redi Gentiluomo Arentino e Accademico della Crusca* (Milano: Società Tipografica de' Classici Italiani, 1811), vol. III, p. 292 [My translation].

27   Cf. Bernardo Fantini, *La rivoluzione pastoriana e le politiche di igiene pubblica*, in Guido Cimino e Bernardo Fantini (ed. by), *Le rivoluzioni nelle scienze della vita* (Firenze: Olschki Editore, 1995).

28   Francesco Redi, *Osservazioni intorno agli animali viventi che si trovano negli animali viventi* (Firenze: Pietro Matini, 1684).

29   William Harvey's *bestiarium*, gathered in his 1628 *Exercitatio anatomica de motu cordis et sanguinis in animalibus*, is as much diversified: here he illustrates the circulatory system of more than forty different species.

Among the possible examples of his bestiary and comparative approach, I have selected one that recalls the Aristotelian passage about animals starved to death for a correct vision on their veins. Here Redi tries to establish both how long animals could survive without either food or water, and to study their inner organs rendered perfect due to starvation.

Gli animali non muoiono cosi prestamente per cagione del digiuno, come crede il volgo. Tra i cani, che ho fatto morir di fame, vi sono stati quelli, che senza mangiare, e senza bere son campati trentaquattro e trentasei giorni. Un piccolo cagnuolo ne' giorni più caldi della estate arrivò fino a venticinque giorni senza bere e senza mangiare; e molto più oltre sarebbe trascorso, se spinto dal gran rovello della fame non fosse saltato da un'altissima finestra. Un gatto del Zibetto [...] indugiò a morire dieci giorni, e un grossissimo gatto selvatico ne indugiò venti. Venti giorni mi campò una gazzella. Un tasso, in tempo d'inverno, campò un mese intero. I topi domestici, e campagnuoli possono poco soffrir la fame; imperocché in molte prove che ne ho fatte, non son mai arrivati a tre giorni interi senza mangiare. Pel contrario le tartarughe terrestri le ho condotte fino a diciotto mesi; le vipere fino a dieci; e come ho detto di sopra, un lucertolone africano campò più di otto mesi [...].

Non e' immaginabile, quanto si trovino belle le viscere degli animali fatti morir di fame; il che dovrebbe servire per insegnamento, che la dieta ben regolata e' la più sicura medicina, per rimettere in sesto le viscere degli uomini, e per stasare gl'intrigatissimi canali, e andirivieni de' loro corpi.[30]

---

30   'Animals do not die as quickly as people think, when they are not fed. Among the dogs I starved to death, there were some who lived thirty-four and thirty-six days without eating and drinking. During the hottest days of the summer, a small dog was able to survive twenty-five days with neither water nor food, but it could have survived longer if the nagging thought of the hunger did not induce him to jump out from a very high window. An African civet [...] survived ten days, and a big wildcat twenty. A gazelle lived twenty days. A badger during the winter lived for a whole month. Domestic and wild rats cannot sustain hunger, therefore in several trials I did, they never reached three entire days without eating. On the contrary terrestrial turtles reached eighteen months; vipers ten; and as I previously said, an African lizard was kept alive for more than eight months [...].
It is not conceivable how beautiful are the intestines of the animals starved to death. This should teach us that a well-regulated diet is the safest remedy to restore the intestines of humans, and to unclog the tangled canals and labyrinth of their bodies.' Francesco Redi, *Osservazioni intorno agli animali viventi*, in

The treatise has features in common with the ancient zoological catalogues, insofar as animal heterogeneity is the texture for the comparativeness of a theoretical, biological framework. Both Aristotle and Redi find that life sciences have to pass through death, and pass through very slowly, in this case.

I want to insist on the element of variety because it will be the cornerstone upon which nineteenth-century experimental biology will change its route. In Redi's works, as in Aristotle's, there is almost no selection of the material to study. Specimens are 'casual animals', mostly wild animals (such as vipers, wildcats, gazelles, turtles, lizards, etc.) or sinantropic ones (such as dogs, rats, doves and pigeons). In any case they are not bred, and Redi does not seem to deploy any kind of conscious choice.

But something has changed from the ancient form of dissection. For Aristotle, the 'operation' of giving death had heuristic value, because it allowed him to see something deeply veiled into the living body. For Redi, anatomical perception is more than heuristic, it is statistic, and so it has to become systematic and repetitive. The reiteration of an observation represents a conspicuous feature of the early modern experiment,[31] and Redi interprets its necessity in terms of an 'iterata e reiterata esperienza'. As he says: 'Every day I am more firm in my intention of not trusting the phenomena of nature if I do not see them with my own eyes and if they are not confirmed by iterated and reiterated experience.'[32]

A golden rule for experimentation thus emerges: reiteration increases the reliability of the trial and, literally, makes natural phenomena believable. Here, something quite different from the Aristotelian, spontaneous manifestation of truth is occurring: nature has to be pushed and provoked.

Opere di Francesco Redi Gentiluomo Arentino e Accademico della Crusca (Milano: Società Tipografica de' Classici Italiani, 1811), vol. III, pp. 291-292, [My translation].

31   Galileo Galilei claimed to have repeated 'a full hundred times' his experiments with the inclined plane as described in *Two New Sciences*. Meanwhile, the motto of the Florentine Accademia del Cimento was 'provando e riprovando,' testing and re-testing, cf. Jutta Schickore, 'Trying Again and Again: Multiple Repetitions in Early Modern Reports of Experiments on Snake Bites', in *Early Science and Medicine*, 15, 2010, pp. 567-617.

32   It is the very beginning of his first work, *Observations on Vipers* (1664), Peter Knoefel (ed. by), *Francesco Redi on Vipers* (Leiden: Brill, 1988), p. 3.

The weight ascribed to the experimental reiteration of death can be further gleaned by reading another passage in which Redi participates in the debate about the effectiveness of vipers' venom after the viper's death.

> At eleven o'clock on the morning of May the ninth, I had the heads cut off of a great number of vipers; an hour and a half later, when they were dead and the heads remained motionless, I took one of those heads in my hand and opening wide its mouth, made it bite a city dove in the breast muscle and I pressed firmly the head, so that that yellow, which dwells in the sheath of the viper's major teeth, could well penetrate into the wound of the bite; the dove died in a little less than two hours.
>
> On May the tenth, thirty-three hours after the death of the vipers, I made another of those heads bite another dove in the breast [...]. The next day, the eleventh of May, fifty-four hours after the viper's death, I had three doves bitten by different heads; the first fell dead almost immediately; in the second, death was delayed for two hours, nearly three hours in the third. [...]
>
> In the middle of June I repeated the previous tests with dead vipers' heads, and always death resulted in animals bitten; however I could continue them for three days only, because in the great heat of the summer the heads reached complete stinking corruption.[33]

Death seems to be simultaneously an easy mechanical operation of ordinary science, and the darkest mirror in the depths of which one can recognize life's movements. But it is still an early modern death, given (from the scientist) through bites and destined to a 'stinking corruption'.

Normalized Animals for Scientific Truth

> Breeders habitually speak of an animal's organization as something quite plastic, which they can model almost as they please
>
> – Charles Darwin

In the second half of the nineteenth century, Louis Pasteur demonstrated his revolutionary immunological theory which showed its great importance not only on scientific and epistemological grounds, but also

---

33   Knoefel (ed. by), pp. 53-54. For an analysis of the same passage of the treatise, see Schickore.

on the social one, as it contributed more than anything else to the complex process of the medicalization of society. Pasteur developed his theory in the context and with the urgency of particular historical circumstances: an epizootic recrudescence and the long series of cattle diseases that raged through Europe.[34] This means that livestock was at once his case study, his experimental material, and the beneficiary of his research.

To understand the development of biological discourse, it is very important to briefly consider the zootechnics' industrial conversion.[35] In the eighteenth century, zootechnics started to undergo a rationalization process: breeders and agronomists living and working in this transitional moment used the expression 'rational breeding' for their new techniques and facilities of rearing.[36] The improvements in chemistry and biology, as well as those in technical equipment, in addition to the demographic and urban growth and its needs, forced breeders to optimize their production. Breeders were ready for that: even in the middle of eighteenth century, they were acutely aware of the possibilities of animal selection[37] as the most effective among procedures of rationalization.

The coexistence and co-operation of zootechnics with the discoveries appearing between the Scientific Revolution and the Industrial Revolution is very complex, however three major areas of interaction can be detected.

A deep interaction with biological sciences begins: from breeders they obtain their living material, and in return they provide theoretical models. But zootechnics begins also a dialogue with social sciences, concerning topics such as docility, or the governmentality of populations of living beings. Sociobiology, and its preparatory role towards eugenic

---

34   Epizootics started to be studied at the end of the seventeenth century, but it was in the nineteenth century, amid the large, crowded farms of European and American suburbs, that they saw a serious revival. Cf. René Dubos, *Pasteur and Modern Science* (New York: Doubleday, 1960); Fantini.

35   Raphael Blanchard, 'Zoologie et médecine', in *Actes du VIᵉ Congrès International de Zoologie* (Berne: 1905).

36   Arthur Young, *Annals of Agriculture, and other Useful Arts* (London: Bury St. Edmund's, 1784-1815); Jean Baptiste Boussingault, *Économie rurale cosidérée dans ses rapports avec la chimie, la physique et la météorologie* (Paris: Libriaire-editeur Béchet Jeune, 1851); Adrien de Gasparin, *Cours d'agriculture* (Paris: Librairie agricole de la maison rustique, 1863).

37   It was Charles Darwin who, in *The Origin of Species by Means of Natural Selection* (London: John Murray, 1859), confesses his debt to the centennial practical knowledges of breeders with regard to the principles of selection, especially pp. 29-37.

policies of the twentieth century are emblematic in this regard. Lastly, animal breeding has a mutual relation with economic and mechanical sciences, in creating the theories of labor rationalization, useful in the context of industrialization and mass production.

These three dimensions seem to reveal the three faces of the industrial animal: animal as *organism*; animal as *subject*; animal as *thing*. These three figures are gathered under the new hegemonic concept of animal as a *product*. Whether the animal is considered a subject (for ethology and social sciences), an organism (for biology), or a thing (for techniques), in the nineteenth century it becomes primarily a product. In contrast with the previous creationist paradigm, in which the animal was the figure representing Creation, livestock animals become the effect and product of, at once, natural history, zootechnic history, and social history. Hence, the systematization of breeding knowledges concerning genetics, such as selection and crossbreeding, plays an important role in the emergence of this new conceptualization of the animal as a product.

Evolutionism also plays its part in this new syntax. As Darwin started to use animal breeding to explain natural history, linking them in a mutual epistemological field, he naturalized zootechnics on the one hand, and artificialized nature on the other. The main concept of evolutionism is that of 'natural selection'. In Darwin's choice of these two otherwise oxymoronic words, we can begin to understand and evaluate the importance of his indebtedness to breeders' knowledge and to its conceptual implications. By speaking of nature through the concepts and categories of zootechnics, Darwin radically transformed the representation of nature itself. Natural history, based on the model of zootechnics, is thus combined with the industrial production to become a colossal factory of living beings. Industrial breeding appeared to Darwin and to his – and our – contemporaries as the rationalized continuation of nature.

The animal represents, at this height, the missing link between matter and *bíos*, inside the industrial complex imaginary. At the boundary that divides persons from things, animals, in a very long tradition, occupied precisely the place of this original separation, but also of their secret commonality.[38] The animated and the inanimate, in the productive process, find their terrifying confusion in the *living-object* of a radical biopolitical regime like that of the industrial farming. If industrial

---

38   Cf. Roberto Esposito, *Persons and Things: From the Body's Point of View*, trans. by Zakiya Hanafi (Cambridge: Polity Books, 2015).

breeding lent disciplinary strategies to modern structures such as the prison, the psychiatric asylum, and finally concentration camps, then it was also one of the greatest sources of inspiration for mechanical industry, as Henry Ford himself states: 'The idea [of the moving assembly line] came in a general way from the overhead trolley that the Chicago packers use in dressing beef.'[39]

For all these reasons, the many connections that familiarize zootechnic institutions with social policies on the one hand, and industrial production techniques on the other, should be traced accurately. But biology has been the most fruitful field for a lexical, epistemological, and material exchange. As the breeding farm came to be configured as an apparatus of profound normalization, it was revealed to be the *conditio sine qua non* for a strict adoption, by biological sciences, of scientific parameters for experiment.

From early modernity, the criteria that allow an experiment to be considered scientific are reiteration of observation, and laboratory determinability (its hermetic closure, and the surveillance that the scientist can ensure). Hence, biology cannot but find some difficulties in being recognized as a proper science, since its 'laboratory' is literally the living organism. Biological sciences have to maintain their ambiguous epistemological status from their inception up until today, in the sense that they mostly use the experimental criteria that they share with physical-chemical sciences, trying to resemble them. But at the same time, their object is irreducible to the model used by physics and chemistry. It is with the highly-selected – and therefore serialized – animals of the new factory farming that experimental biology finds a way of negotiating its compromise with physics and chemistry.

If we read Pasteur's accounts of his experiments we find that he never worked on a single animal, but always with stocks of animals, considered as identical and exchangeable. He was searching for closed regular systems and therefore he needed to reduce variants to a fixed model, upon which the theory can be abstracted. Here is a representative passage of his method:

> I take eighty new chickens (I call *new* those which never suffered before with chicken cholera). Twenty of these I inoculate with the most virulent virus, and they all die. Of the sixty that remain, I take another lot of twenty, and I inoculate them with that quantity of the most

---

39   Henry Ford, *My Life and work* (New York: Garden City, 1922), p. 81.

attenuated virus which the point of the needle will take up – and not one dies. Are they then vaccinated for the aggravated form of virus? Some are and some are not, for if I afterwards inoculate these twenty chickens with the most virulent virus, six or eight of them will not die, although they may be ill, while in the first case every inoculated chicken died. I take again from the remaining chickens another lot of twenty, and these are vaccinated with the attenuated virus exactly as the preceding lot, and, a week afterwards, they are again vaccinated in the same manner. Are they now safe from the virulent virus? We now inoculate these twenty chickens with this virulent virus, and, instead of there being six or eight which do not die, there are twelve or fifteen. Finally, I take the twenty remaining chickens, and vaccinate them successively three or four times. If now I come to inoculate them with the most virulent virus, not one will die. In this case, chickens are brought to the condition of animals which are incapable of suffering from chicken cholera.[40]

If seventeenth-century scientists worked substantially with the 'casual material' of captured, wild or sinantropic animals, nineteenth-century scientists work otherwise with livestock. This is a meaningful difference because animals from the new factory farming are highly selected and serialized specimens. In a (Foucauldian) word: they are *normalized* animals.[41]

In this *normal animal* scientists seem to discover 'bare life'.[42] Science needs to strip reality of its contingency, so that it can reach the mere fact, and its purity. The *veridiction*[43] required by science introduces a radical dichotomy between *fact* and *fetish*.[44] The scientific fact is naked and pre-exists human intervention. By contrast, the fetish is a product, an effect of human intervention. This is the reason why fact is the privileged

---

40 Louis Pasteur, 'On Chicken Cholera: Study of the Conditions of Non-Recidivation and of Some Other Characteristics of this Disease', trans. by Paul Casamajor, *Science*, vol. 2, n. 32, 1881, 55-57, p. 56.
41 Bruno Bianchini, *L'evoluzione della zootecnia attraverso i secoli* (Terni: Stabilimenti Poligrafici Alterocca, 1930).
42 Giorgio Agamben, *Homo Sacer: Sovereign Power and Bare Life*, trans. by Daniel Heller-Roazen (Stanford: Stanford University Press, 1998).
43 That of *Veridiction* is a concept coined by Michel Foucault to point at an authority's faculty to decide what is to be considered true and false. Each epoch has its 'regime of truth', and science is the particular regime of truth of the contemporary era.
44 Cf. Bruno Latour, *On the Modern Cult of the Factish Gods*, trans. by Heather MacLean and Cathy Porter (Durham, NC: Duke University Press, 2009).

interlocutor of science, because it tells humans something that does not derive from them, and that they therefore do not know. But because the scientific factuality never corresponds to the phenomenal one, it must be pushed and provoked or, more precisely, artificially created through a process of radical denudation of the fact from its optional accidents. Pointing to the high level of intervention required for the experiment to actually be scientific, Latour calls scientific objects *factish*, which Western scientists are well-disposed to listen to and to venerate. In this sense, the animal from factory farming is precisely an oracle in this sense. Despite – or rather *because* – of the fact that living being is the product of a radical biopolitical regime, the zootechnic specimen is chosen as scientific fact.[45] Here we find the device activated by biological sciences in order to lay scientific foundations for the study of living beings. Since nature, chaotic and phenomenal, does not spontaneously offer up the regularity and normality required by the experiment, scientists are forced to produce a second nature, a *para-nature* with which to replace the first one.

If in ancient epistemology, retraceable from Aristotle's researches, animal *physis* manifests its own truth, in Redi's modern epistemology, *natural* and *biological phenomena* must be tested and re-tested in order to become true. In Pasteur's almost contemporary epistemology, finally, *experimental animals* substitute natural ones as the only ones who can pronounce scientific truth.[46]

In conclusion, my thesis is that animals represent one of the focal points where, in the most literal and glaring way, certain forms of knowledge set up from certain techniques of power.

---

45   'The naturalistic animal provides the conditions for achieving its analytic counterpart', Michael Lynch, 'Sacrifice and the Transformation of the Animal Body into a Scientific Object: Laboratory Culture and Ritual Practice in the Neurosciences', *Social Studies of Science*, 18, 1988, p. 280. Also cf. Massimo Petrozzi, *Touching the Body of the Animal: Studying Animals in Early Modern Florence, 1650-1700* (Baltimore: 2010).

46   Guerrini says it otherwise: 'The role of animals in anatomy and natural history changed over the course of the sixteenth and early seventeenth centuries from passive objects of observation to active subjects in demonstration and experimentation. […] dissection of live animals moved beyond demonstration to become a quintessentially experimental act, an act that revealed new knowledge.' (Guerrini, pp. 7-10). This process is part of the increasing constructiveness of modern science, that uses animal dissection to build scientific truth (more than simply showing it), and simultaneously needs already artificial animals.

If, as shown by historical epistemology, sciences proceed from techniques rather than the contrary, this is evidently the case of biology. The science of the living, indeed, developed its specific epistemological frame from the material and symbolic construction of its objects of study, the animals. Construction that has been carried out by the techniques devoted to animal manipulation: firstly by the practices of hunting, and afterwards by the zootechnic apparatuses. These are a particular kind of techniques, though: insofar they are exercised on living beings they are (also) techniques of *government*, techniques of *power*.

How, in the livestock farm, technology intended as mere manipulation of things – animal bodies in this case – makes way to the government of subjectivities and of populations, therefore to politics? And how both these strategies of management (of things and of living beings) produce a science with an epistemological status so particular, as biology has, and with a political destiny so problematic, as it has in our biopolitical era?

By retracing few fundamental examples of the interrelation between *bio-logy* and *zoo-techniques*, these questions arise. Answering them will be possible only after having filled the gap that currently exists in the history of zootechnics, and developing a critical theory that sets out from it.[47]

Bibliography

Agamben, Giorgio, *Homo Sacer: Sovereign Power and Bare Life*, trans. by Daniel Heller-Roazen (Stanford: Stanford University Press, 1998).

Aristotle, *History of Animals*, trans. by D'Arcy Wentworth Thompson (Oxford: Clarendon Press, 1910).

Bianchini, Bruno, *L'evoluzione della zootecnia attraverso i secoli* (Terni: Stabilimenti Poligrafici Alterocca, 1930).

Bichat, Marie François Xavier, *General Anatomy Applied to Physiology and Medicine*, trans. by Constant Coffyn (London: King's College, 1824).

---

47   In the past few years some attempts in the sense of a 'zoohistoire', as Robert Delort calls it, have been made, even if a systematic survey on modern zootechnics is still missing. The subject promises the opening of important new fields of research, and unpredictable feedbacks on classical historical studies, as Dorothee Brantz states: 'Incorporating animals into any historical narrative, necessitates a radical rethinking of the project of history', Dorothee Brantz (ed. by), *Beastly Natures: Animals, Humans, and the Study of History* (Charlottesville: University of Virginia Press, 2010), p. 3. Cf. also Robert Delort, *Les Animaux ont une histoire* (Paris: Seuil, 1984).

Brantz, Dorothee (ed. by), *Beastly Natures: Animals, Humans, and the Study of History* (Charlottesville: University of Virginia Press, 2010)

Canguilhem, Georges, *Œuvres complètes, Tome I, Écrits philosophiques et politiques (1926-1939)* (Paris: Vrin, 2011).

Canguilhem, Georges, *Knowledge of Life*, trans. by Stefanos Geroulanos and Daniela Ginsburg, (New York: Fordham University Press, 2008).

Canguilhem, Georges, *Ideology and Rationality in the History of the Life Sciences*, trans. by Arthur Goldhammer, (Cambridge: MIT Press, 1988).

Carbone, Andrea L., *Aristote illustré. Représentatio du corps e schématisation dans la biologie aristotélicienne* (Paris: Classiques Garnier, 2011).

Cimino, Guido, and Bernardo Fantini (ed. by), *Le rivoluzioni nelle scienze della vita* (Firenze: Olschki Editore, 1995).

Cohen-Rosenfield, Leonora, *From Beast-Machine to Man-Machine: The Theme of Animal Soul in French Letters from Descartes to La Mettrie* (New York: Octagon Books, 1968).

Cunningham, Andrew, *The Anatomical Renaissance: the Resurrection of the Anatomical Projects of the Ancients* (Aldershot: Scolar Press, 1997).

Darwin, Charles, *The Origin of Species by Means of Natural Selection* (London: John Murray, 1859).

Delort, Robert, *Les Animaux ont une histoire* (Paris: Seuil, 1984).

Didi-Huberman, Georges, *Ouvrir Vénus. Nudité, rêve, cruauté* (Paris: Gallimard, Parution, 1999).

Dubos, René, *Pasteur and Modern Science* (New York: Doubleday, 1960).

Esposito, Roberto, *Persons and Things: From the Body's Point of View*, trans. by Zakiya Hanafi (Cambridge: Polity Books, 2015).

Ford, Henry, *My Life and Work* (New York: Garden City, 1922).

Foucault, Michel, *The Birth of the Clinic: An Archaeology of Medical Perception*, trans. by Alan Sheridan (London: Routledge, 2003).

Foucault, Michel, *The Order of Things: Archaeology of the Human Sciences*, trans. by Alan Sheridan (New York: Vintage, 1973).

French, Roger, *Dissection and Vivisection in the European Renaissance* (Aldershot: Ashgate, 1999).

Fudge, Erica (ed. by), *Renaissance Beasts: of Animals, Humans, and Other Wonderful Creatures* (Urbana and Chicago: University of Illinois Press, 2004).

Guerrini, Anita, *The Courtiers' Anatomists: Animals and Humans in Louis XIV's Paris* (Chicago: The University of Chicago Press, 2015).

Knoefel, Peter K. (ed. by), *Francesco Redi on Vipers* (Leiden: Brill, 1988).

Latour, Bruno, *On the Modern Cult of the Factish Gods*, trans. by Heather MacLean and Cathy Porter (Durham: Duke University Press, 2009).

de' Liuzzi, Mondino, *Anothomia*, trans. by Piero Giorgi and Gian Franco Pasini (Bologna: Istituto per la storia dell'Università di Bologna, 1992).

Pasteur, Louis, 'On Chicken Cholera: Study of the Conditions of Non-Recidivation and of Some Other Characteristics of this Disease', trans. by Paul Casamajor, Science, vol. 2, n. 32, 1881, pp. 55-57.

Pellegrin, Pierre, *La classification des animaux chez Aristote* (Paris: Les Belles Lettres, 1982).

Petrozzi, Massimo, *Touching the Body of the Animal: Studying Animals in Early Modern Florence, 1650-1700* (Baltimore, 2010).

Redi, Francesco, *Osservazioni intorno agli animali viventi*, in Francesco Redi, *Opere di Francesco Redi Gentiluomo Arentino e Accademico della Crusca* (Milano: Società Tipografica de' Classici Italiani, 1811).

Ruini, Carlo, *Anatomia del cavallo infermità et suoi rimedii* (Venice: Fioravante Prati, 1618).

Schickore, Jutta, 'Trying Again and Again: Multiple Repetitions in Early Modern Reports of Experiments on Snake Bites', in *Early Science and Medicine*, 15, 2010.

Vegetti, Mario, *Il coltello e lo stilo. Le origini della scienza occidentale* (Milano: il Saggiatore, 1996).

Vesalius, Andreas, *On the Fabric of the Human Body*, trans. by William Frank Richardson and John Burd Carman, 5 vols (San Francisco and Novato: Norman Publishing, 1998-2009).

Zucker, Arnaud, *Aristote et les classifications zoologiques* (Paris: Peeters, 2005).

PETER KOFLER

THE ANACHRONOUS MONTAGE
OF MAN AND SERPENT

What does actually happen, if a man takes a poisonous serpent into his mouth? This is, among a variety of other topics and reflections, what Aby Warburg describes in his famous *Lecture on Serpent Ritual*, held on April 21, 1923 during his stay at Ludwig Binswanger's Bellevue clinic in Kreuzlingen.

As is widely known, during his journey through North America at the end of the 19th Century Warburg had the opportunity to observe some of the rites of the Native Americans, inspired by the American anthropologist and ethnologist Frank Hamilton Cushing, a pioneer in the studies of the Zuni Indians of New Mexico. In his Kreuzlingen lecture Warburg reports Cushing's account

> of what an Indian had once said to him: 'Why should a man stand higher than an animal? Look at the antelope. It is a Run. It runs so much better than a man. Or the bear, it is just Strength. Men can only do a little; an animal can do wholly what it has in it to do.'[1]

What Cushing indirectly refers to is the antelope-dance in San Ildefonso, of which Warburg took a series of pictures in 1896. But in wider terms, the statement is a claim for the subversion of the hierarchy existing between man and animal. Animals are superior to men thanks to their ability to fully accomplish what nature has destined them for, unlike men, who, in spite of their many skills, are unable to fully exploit each of them. 'For the Indian', Warburg states, 'the animal is a higher being, because the integrity of the animal-nature makes it seem a more gifted creature in contrast to man, who is weak.'[2]

---

1   Aby Warburg and W. F. Mainland, 'A Lecture in Serpent Ritual', *Journal of the Warburg Institute*, vol. 2, no. 4 (Apr., 1939), 227-292 (p. 283).
2   Ibid.

If on the one hand animals are considered physically superior to men, on the other hand they appear to be endowed with something which is usually attributed only to humans. Warburg argues that a superstition leads the Pueblo Indians to believe that animals, and even plants, are provided with 'active souls'[3] and that this accounts for 'unlimited possibilities of relationship between man and his environment.'[4] Seeking to reestablish their lost perfection, men developed magical practices through which they tried not only to imitate animal behavior, but, literally, to transform themselves into animals for the short time of a mimetic dance. The goal of such a performance is essentially a practical one, as Warburg explains: 'By slipping into the animal mask in the hunting-dance the animal is captured in anticipation by a miming of the attack to be made during the actual hunt.'[5] But apart from that there is also an ontological dimension which comes into question. Warburg continues:

> In establishing contact with something entirely non-personal the masked dance means to primitive man the most profound submission to some external being. When the Indian in his miming costume imitates for example an animal by movement and sounds, he is trying to transform his own self and so to wrest from nature by magic means something to which he feels he cannot attain so long as his personality remains unchanged and unextended [6]

Rather than with an example of pure mimesis, what is at stake here is a real metamorphosis or, more precisely, metamorphosis through mimesis. Although he confesses that he has not personally attended to it, Warburg is convinced that '[t]he supreme example of magical assimilation to nature by way of the animal-worlds is found among the Moki Indians in their dances with live serpents at Oraibi and Walpi.'[7] And the main reason is that, as Warburg points out, 'here the dancers and the live beast form a magic unity.'[8] Furthermore, what marks the

---

3   'aktive Seelen'. Aby Warburg, *Schlangenritual: Ein Reisebericht* (Berlin: Klaus Wagenbach, 1988), 13. Translation by author.
4   'ein befreiendes Erlebnis der schrankenlosen Beziehungsmöglichkeit zwischen Mensch und Umwelt'. Ibid. Translation by author.
5   Warburg and Mainland, p. 282.
6   Ibid.
7   Ibid., p. 286.
8   Ibid.

difference between the brutal and sanguinary character of other Indian rites and the Moki ones, is the fact that they 'have a way of handling the most dangerous of reptiles, the rattle-snake, so that it can be tamed without violence and will join the ceremonies for days on end with complete docility'.[9]

Besides this surprising non-violent property of the Moki ritual, to which I'll get back soon, it is particularly interesting to note that for Warburg the transformation of man into animal undergoes three distinct stages. In one of the sketches for the Kreuzlingen lecture Warburg outlines his concept of assimilation (*Einverleibung*), distinguishing among *Verleibung*, *Anverleibung*, and self-loss.[10] *Verleibung* involves a kind of interaction through which an object is completely absorbed by the subject; *Anverleibung* indicates that parts of the object remain visible as foreign material expanding the self; and, finally, with self-loss, the complete subjugation of the self to an object is reached.[11] When the Moki dancer bears and handles a live serpent, he gives way to an extension of his self, called *Anverleibung*; and when he takes it into his mouth, he achieves what Warburg calls *Verleibung* or *Einverleibung*, through which, as we have seen earlier, a magic unity of man and animal is performed, and, consequently, the difference between the two is blurred, if not altogether cancelled. Humans melt with or turn back to a primitive level of animality. At this point, Darwinism obviously lies just around the corner. And indeedt, in his lecture Warburg refers to the theory of the famous English naturalist:

> In a state of reverential awe, in what is called totemism, these pagan Indians, like all the pagans throughout the world, unite with the animal kingdom by believing in animals of all kinds as mythical ancestors of their tribes. Their explanation of nature by imaginatively interrelating man with animals is not so very far removed from Darwinism; for as we impute a physical law to the process of evolution in nature, the pagans try to establish an imaginary association of man with the animal-world. The decisive factor in the lives of these so-called primitive peoples may be called a kind of mythical Darwinism of elective affinities.[12]

---

9   Ibid.
10 Aby Warburg, *Werke in einem Band*, ed. by Martin Treml, Sigrid Weigel, and Perdita Ladwig (Berlin: Suhrkamp, 2010), p. 590.
11 Ibid.
12 Warburg and Mainland, p. 283.

In this passage Warburg does not refer so much to *The Origin of Species* (1859), where Darwin didn't dwell particularly on the affinity between man and animal,[13] but rather to two of his later works: *The Expression of the Emotions in Man and Animal* (1872) and *The Descent of Man* (1874), where, as Markus Wild points out, Darwin 'establishes that there is no fundamental difference between man and the higher developed animals in terms of their emotional, mental and moral abilities.'[14] However, the text that most influenced Warburg was *The Expression of the Emotions in Man and Animal*, which he read as early as 1888 at the Florence National Library.[15] Starting from these premises, we cannot seriously expect Warburg's reading of Darwinism to be in line with the idea of the evolution of species. It is, rather, a reversal, or, better, a sort of overlapping, a biological palimpsest allowing for a double reading, which puts together, like in a constellation, two different evolutionary epochs, or, in the words of Georges Didi-Huberman, an 'anachronous montage'[16] of man and animal.

The same kind of anachronism and chronological juxtaposition of different evolutionary stages, can be found in art history, too. And surely the most famous and fitting example is represented by the marble group *Laocoön and his Sons*, attributed by Pliny the Elder to the Rhodian sculptors Agesander, Athenodoros, and Polydorus and depicting the struggle of the Trojan priest and his sons against two deathly serpents. Didi-Huberman writes: 'Who would doubt that *Laocoön*, the famous masterpiece of Greek sculpture, updated a primitive cultural condition

---

13   Markus Wild, 'Anthropologische Differenz', *Tiere: Kulturwissenschaftliches Handbuch*, ed. by Roland Borgards (Stuttgart: Metzler, 2016), 47-59 (p. 53).

14   'führt Darwin den Nachweis, dass es keinen fundamentalen Unterschied zwischen dem Menschen und den höheren Tieren hinsichtlich ihrer emotionalen, mentalen und moralischen Vermögen gebe'. Ibid. Translation by author. See also Eve-Marie Engels, 'Darwin/Darwinismus', *Lexikon der Mensch-Tier-Beziehungen*, ed. by Arianna Ferrari and Klaus Petrus (Bielefeld: transcript, 2015), 69-73 (p. 71).

15   Georges Didi-Huberman, *Das Nachleben der Bilder: Kunstgeschichte und Phantomzeit nach Aby Warburg*, trans. by Michael Bischoff (Berlin: Suhrkamp, 2010), p. 265; see also Gombrich, Ernst H., *Aby Warburg: Eine intellektuelle Biographie* (Hamburg: Europäische Verlagsanstalt, 1981), p. 99 and p. 327.

16   On Warburg's theory of 'afterlife' ('Nachleben'), Didi-Huberman writes: Zur Theorie des 'Nachlebens' bei Warburg schreibt Didi-Huberman: [...] 'das Verhältnis zwischen Aktualität und Primitivität ist durch eine *anachrone Montage* bestimmt'. Didi-Huberman, p. 248.

of tragic pathos by transformation and disguise?'[17] And he goes even further, wondering if Laocoön's pain could not actually be regarded as the visual representation of an even more ancient relationship bringing together 'the human body and both the physical affliction and the violence of animal fight' and rooted in a pre-symbolic and pre-narrative time.[18] As in the case of the serpent ritual, what we are dealing with here is, again, the contiguousness of the human and the animal.[19]

In the essay on Botticelli written in 1898, two years after his American experience, Warburg speaks of the *Laocoön Group* in terms of *exemplum doloris*, calling it a 'masterly representation of a threefold pain'.[20] Sixteen years later, in a lecture entitled *Der Eintritt des antikisierenden Idealstils in die Malerei der Frührenaissance*, referring to a copy of the *Lacoön* discovered by Luigi Lotti in 1488, Warburg highlights the Janus profile of this 'pathos formula' by emphatically stressing the fact that 'classical perturbation'[21] – in plain contrast to Winckelmann's well known and at that time still authoritative neoclassical principle of 'noble simplicity and quiet grandeur' – was part of the ancient heritage just as its antipole and that, thus, Apollonian ethos and Dionysian pathos were inextricably combined 'in the symbol of a 'double herm'.[22]

As a result, it could be assumed that also the serpent has a double nature, grounded on what the ancient Greeks called *pharmakon*. Following Plato's *Phaidros*, Jacques Derrida elucidates the meaning of the word, which refers to both remedy and poison, not in the sense that it is endowed with antithetic effects, but exactly because it has no special

---

17 'Wer wollte bezweifeln, daß der *Laokoon*, dieses Meisterwerk der griechischen Plastik, einen kulturell primitiven Zustand des tragischen Pathos aktualisierte – indem er ihn verwandelte und verkleidete?'. Didi-Huberman, p. 249. [Translation by author]
18 'Verhältnis [...] des menschlichen Körpers zum physischen Leiden und zur Gewalt des animalischen Kampfes'. Ibid. Translation by author.
19 Ibid.
20 'die meisterhafte plastische Verkörperung dreifachen Schmerzes'. Aby Warburg, 'Sandro Botticelli', A.M.W., *Die Erneuerung der heidnischen Antike: Kulturwissenschaftliche Beiträge zur Geschichte der europäischen Renaissance*, ed. by Horst Bredekamp and Michael Diers (Berlin: Akademie Verlag, 1998), 61-68 (p. 68). Translation by author.
21 'klassische Unruhe'. Aby M. Warburg, 'Der Eintritt des antikisierenden Idealstils in die Malerei der Frührenaissance', A.M.W., *Die Erneuerung der heidnischen Antike: Kulturwissenschaftliche Beiträge zur Geschichte der europäischen Renaissance*, ed. by Horst Bredekamp and Michael Diers (Berlin: Akademie Verlag, 1998), 173-176 (p. 176). Translation by author.
22 'im Symbol einer 'Doppelherme''. Ibid. Translation by author.

properties of its own. For Derrida, the *pharmakon*, being deprived of any identity, is 'aneidetic', i.e. anterior to any kind of differentiation.[23]

Thus, starting from Warburg's perspective, human history can be reinterpreted in terms of a gradual transformation of the symbol of the serpent in its trajectory from ancient East to contemporary North America. In his 1923 lecture, Warburg states: 'The emancipation from blood-sacrifice is an ideal of purification which has left a profound mark on the development of religion from east to west. The snake too undergoes this process of sublimation.'[24] It is a slow movement from predominant poisonousness to prevailing healthfulness, the former being exemplified by the *Laocoön Group*, the latter by the Moki dance. What we are facing in the first case is the violent assimilation of man by serpents, while in the second one we can see its reversal, i.e. the magical assimilation of the serpent by man.

If the 'conception of the snake as a destroying power from the nether-world has found its most moving tragic symbol in the myth and in the sculptured group of Laocoon',[25] there is nevertheless a highly prominent example of a serpent symbol in classical antiquity which was intended as a remedy. In Warburg's words:

> [The] snake as a demon expressive of ancient pessimism finds its counterpart in an antique snake-god in which we can at last recognize the benevolence and transfigured beauty of the classical age. Asclepius, the god of healing, has a snake twined round his staff as a symbol [...][26]

Moreover, even in this case we come across a unity of man and serpent, for the 'snake twined round his staff and he himself are one and the same – a departed soul that goes on living and reappears in the form of a serpent.'[27] On the other hand, in his Kreuzlingen lesson Warburg calls his audience's attention on the violent character of most Native Americans' rites.

Turning back to the relationship between man and animal, we will find Didi-Huberman underscoring the claim that it was Darwin who drove Warburg to consider 'the pathos formulae of *Laocoön* [...] in

---

23   Jacques Derrida, *La farmacia di Platone*, trans. by Silvano Petrosino (Milano: Jaca Book, 1985), p. 109.
24   Warburg and Mainland, p. 288.
25   Ibid.
26   Ibid.
27   Ibid., p. 289.

terms of survival of the primitive'[28] and to take Laocoöns gestures as a 'residuum of primitive physical reactions.'[29]

This is exactly the point Balaji Mundkur focuses on in his stimulating interdisciplinary investigation on *The Cult of the Serpent*. What the author contends is 'that long before ophidophobia was vented in cultural traditions, certain ecological forces aided by the chemistry of the body left an indelible mark upon the human mind.'[30] Awe of the serpent's sinuous movements, Mundkur suggests, 'was fixed in man's psyche during anthropogenesis and is reflected in extraordinary ways in his animal behavior, which is inseparable from his social behavior and religious beliefs involving the serpent.'[31] Thus, ophidophobia, being 'traceable to monkeys and anthropoid apes',[32] is one of the central links between man and animal. In fact, the main hypothesis of the study is 'that unlike almost all other animals, serpents, in varying degree, provoke certain characteristically intuitive, irrational, phobic responses in human and nonhuman primates alike' and 'that, in this respect, the distinctiveness of man as a 'logic'-employing, symbol-devising species is blurred'.[33]

On its long and varying way from East to West, the ambivalent pathos formula of the serpent re-emerges in one of the most remarkable and at the same time enigmatic images of Romanesque art. In the lower part of the apse of St. Jakob in Kastelaz near Tramin (South Tyrol, Italy) we can see a fresco dating back to the beginning of the 13th Century showing a series of hybrid creatures. Most scholars suggest that it could be a bestiary, an allegorical representation of 'unknown demonic forces',[34] a

---

28   'Er betrachtete die 'Pathosformeln' des Laokoon [...] aus der Sicht eines Nachlebens des Primitiven'. Didi-Huberman, p. 259. Translation by author.

29   'eines [...] Residuums primitiver körperlicher Reaktionen'. Ibid. Translation by author.

30   Balaji Mundkur, *The Cult of the Serpent: An Interdisciplinary Survey of Its Manifestations and Origins* (Albany: State University of New York, 1983), p. 209.

31   Ibid., p. XVI.

32   Ibid., p. 8.

33   Ibid., p. 6.

34   'unbekannter dämonischer Mächte'. Josef Garber, *Die romanischen Wandgemälde Tirols* (Wien: Krystall, 1928), p. 88. Translation by author.

medieval confessional,[35] or a painted sermon for illiterates,[36] in which the main figures stand probably for the eight deadly sins.[37] Recently, Ursula Düriegl has pointed to the apotropaic function of these monsters, in the literal sense of *monstrare*, that is having the power to banish the anxieties of the former believers thanks to their visibility.[38] Rejecting most of these interpretations, Gerd Reichardt has recently hypothesized that the depicted creatures of St. Jakob, far from being allegorical representations of vices, stand for the foreign and the unknown, and, what is most important, 'that those fantastically shaped creatures belong to men and therefore can't simply be associated with a bestiary, a collection of exotic animals.'[39] In the Middle Ages, creatural hybridity, or, in other words, *mirabilia* and *portenta*, were regarded as 'part of the divine redemption plan'[40] and at the same time as part of the real world.[41]

If we take a look at the images of St. Jakob church, we can see a bird-fish woman, a centaur, a fish man and a head on legs on the left part of the apse, and a fin-footed cynocephalus, a two-tailed mermaid, a dolphin rider and a sciapod on the right part. Among the literary sources of these figures there are the ancient Greek physician Ctesias' Indian travel accounts *Indica* (5th century BC), Pliny's *Naturalis historia*, Julius Solinus' *Collectanea rerum mirabilium* (3rd century BC), the well-known *Physiologus* (2nd century), the *Etymologiae* compiled by Isidore of Seville at the beginning of the 7th century, Hrabanus Maurus' *De rerum naturis* (7th century), and the *Imago mundi* by the famous

---

35   Wolfgang Metzger 'Die Apsisfresken von St. Jakob in Kastelatz', *Der Schlern*, 44 (1970), 200-222 (p. 203); see also Franz Dietheuer, 'Das Programm der Apsisfresken in St. Jakob zu Kastelaz ob Tramin', *Der Schlern*, 65 (1991), 307-330 (p. 326).

36   Dietheuer, p. 307.

37   Ibid., p. 318.

38   Ursula Düriegl, *Die Fabelwesen von St. Jakob in Kastelaz bei Tramin: Romanische Bilderwelt antiken und vorantiken Ursprungs* (Wien, Köln, Weimar: Böhlau, 2003), p. 119.

39   'daß diese wunderlich gestalteten Wesen zu den Menschen gehören und damit nicht so einfach einem Bestiarium, einer Sammlung exotischer Tiere, zuzuordnen sind'. Reichardt, Gerd, 'De homine et portentis': Die Apsismalereien von St. Jakob bei Tramin', *Archäologie der Phantasie: Vom 'Imaginationsraum Südtirol' zur 'longue durée' einer 'Kultur der Phantasmen' und ihrer Wiederkehr in der Kunst der Gegenwart*, ed. by Elmar Locher, and Hans Jürgen Scheuer (Bozen: sturzflüge, 2012), 97-109 (p. 104). Translation by author.

40   'Teil des göttlichen Erlösungsplanes' Ibid., p. 106. Translation by author.

41   Ibid.

12th century Christian theologian Honorius of Autun.[42] The unknown painters of St. Jakob must have been familiar with this tradition. Nevertheless, due to the considerable departures from that tradition, the meaning of the fresco remains largely obscure.[43]

Looking at it more in detail, we discover that the bird-fish woman is bearing a scary serpent, and she is brandishing it against her antagonists like a sword or lance. Wolfgang Metzger looks at her in terms of allurement, believing her to be almost certainly an allegorical representation of '(degenerate) *dialectic*',[44] whereas in the opinion of Dietheuer she was, in the eyes of those times, 'the vain antichrist'[45] and, insofar, she stood for vanity.[46] According to this, the serpent she is handling represented the devil.[47] Düriegl holds the bird-fish woman of St. Jakob is identical to a fabulous Indian animal named Mantichora, which Aristotles describes according to the already mentioned Ctesias.[48] The word 'Mantichora' means cannibal in Greek and is related to the Indian tiger, whose voice, according to Honorius, 'is similar to the fizzle of a serpent'.[49] Anyway, what we see here is a kind of synthesis between man and animal. The serpent bearing bird-fish woman represents the type of assimilation Warburg calls *Anverleibung*, the serpent being an extension of the woman's power, as proved by her deadly exhalation.

As we can see, the cynocephalos on the right part of the apse is holding a serpent, too. But, unlike the hybrid snake-bearer at the left, he takes it into his mouth, trying to devour it, while the serpent bites his left shoulder. In this case, we have what Warburg calls *Verleibung* or *Einverleibung*, even if we don't know if it is the serpent who absorbs the cynocephalus or vice versa.

Looking at them from a certain distance, the two figures, put together, are in a way similar to the marble Laocoön, the bird-fish woman playing the role of the vengeful god sending two venomous serpents –

42 Düriegl, pp. 38-39; see also Uli Beleffi Sotriffer, 'Im Umfeld spätromanischer Apsismalerei, St. Jakob/Kastelaz in Tramin und vergleichbare Bildausstattungen', Zeitschrift für Schweizerische Archäologie und Kunstgeschichte, 53 (1996), 89-104 (p. 96).

43 Beleffi Sotriffer, p. 96.

44 '(entartete) Dialektik'. Metzger, p. 211. Translation by author.

45 'der eitle Antichrist'. Dietheuer, p. 320. Translation by author.

46 Ibid., p. 321.

47 Ibid.

48 Düriegl, p. 46.

49 'dem Zischen einer Schlange ähnlich sei'. Ibid., p. 48. Translation by author.

as Dietheuer points out, the double twined tail means plural[50] – and the cynocephalus playing that of the victim. But the scene is also akin to the Moki dance. Didi-Huberman confirms this resemblance: 'It is easy to recognize that the *contiguousness of the human and the animal* is one of Laocoön's main issues, but also of the Indian ritual analyzed by Warburg.'[51] In both cases properties are transferred from the animal to the human: the serpent's sinuous movement is transferred to man. In the case of *Laocoön and his Sons*, 'the serpents seem almost to be an 'over-musculature' of the three figures or a phantasmagorical extroversion of their entrails.'[52] The pictures of St. Jakob offer two examples of primitive struggle. And, finally, the Moki ritual, is aimed at controlling the overwhelming powers of nature by magic.

But the Dionysian sinuousness and poisonousness applies not only to the three examples we already dealt with. It touches also the text of Warburg's lecture and its author. As it is widely known Warburg was affected by a very serious mental disease. Some weeks before his lecture, his state of health was gradually improving, although he was still suffering from panic attacks, which made his body struggle and shake. On March 10, 1923, Ludwig Binswanger's clinical record says that the patient is still 'disturbed, raging, and violent.'[53] On April 4, the Swiss psychiatrist and founder of the Bellevue Sanatorium adds that the patient has 'no deeper awareness of his illness.'[54] A few weeks later, talking about the Kreuzlingen lecture, Binswanger reports that Warburg's illness manifested itself in a sort of confabulation, a rather disorganized commentary, performed with a broken voice, on a series of photographs he had taken during his journey among the American Indians.[55] The same

---

50   Dietheuer, p. 320.
51   'Wie man leicht erkennt, bildet die *Nähe zwischen dem menschlichen und dem Tierischen* ein wesentliches Motiv des *Laokoon*, aber auch des von Warburg studierten indianischen Rituals'. Didi-Huberman, p. 249. Translation by author.
52   'erscheinen die Schlangen fast wie eine 'Übermuskulatur' der drei Figuren oder wie Eingeweide, die in einer Art phantasmatischer Umkehrung von innen nach außen gestülpt sind'. Ibid., p. 250. Translation by author.
53   'agitato, furioso e manesco'. Ludwig Binswanger and Aby Warburg, *La guarigione infinita: Storia clinica di Aby Warburg*, ed. by Davide Stimilli, trans. by Chantal Marazia and Davide Stimilli (Vicenza: Neri Pozza, 2005), p. 123. Translation by author.
54   'Nessuna coscienza approfondita della malattia.' Ibid., p. 124. Translation by author.
55   Ibid., p. 125-126.

applies to his writing. Again, it is Didi-Huberman who calls attention to Warburg's hand-writing in those times: 'The writing suddenly breaks off, being interrupted by nervous or grotesque squiggles or scraggy spirals and, as this is occurring more frequently in 1921-1922, by lightning-shaped lines running all across the page.'[56] Thus, the structure of the text takes on both the shape of the lecturer and that of the subject it is treating of. Warburg's *Lecture on Serpent Ritual* proves to be itself an extremely ambivalent, changing, and protean text. Didi-Huberman is convinced that Warburg was 'aware of the fact that what he was delivering by showing the serpent caught between the Walpi dancers' teeth was an analogy of his own situation'.[57] Thinking about the serpent has its counterpart in a serpent like thinking, expressing itself in body, voice, and text.

Bibliography

Agamben, Giorgio, *L'aperto: L'uomo e l'animale* (Torino: Bollati Boringhieri, 2002).

Beleffi Sotriffer, Uli, 'Im Umfeld spätromanischer Apsismalerei, St. Jakob/ Kastelaz in Tramin und vergleichbare Bildausstattungen', *Zeitschrift für Schweizerische Archäologie und Kunstgeschichte*, 53 (1996), 89-104.

Bender, Cora, Thomas Hensel, and Erhard Schüttpelz (Eds.), *Schlangenritual: Der Transfer der Wissensformen vom Tsu'ti'kive der Hopi bis zu Aby Warburgs Kreuzlinger Vortrag* (Berlin: Akademie Verlag, 2007).

Binswanger, Ludwig and Aby Warburg, *La guarigione infinita: Storia clinica di Aby Warburg*, ed. by Davide Stimilli, trans. by Chantal Marazia and Davide Stimilli (Vicenza: Neri Pozza, 2005).

Cestelli Guidi, Benedetta, and Nicholas Mann (Eds.), *Photographs at the Frontier: Aby Warburg in America 1895-1896* (London: The Warburg Institute, 1998).

Derrida, Jacques, *La farmacia di Platone*, trans. by Silvano Petrosino (Milano: Jaca Book, 1985).

Derrida, Jacques, *L'animale che dunque sono*, ed. by Marie-Louise Mallet,

---

56 'Der Schriftfluß bricht plötzlich ab und wird von nervösen oder grotesken Schnörkeln oder wirren Spiralen durchbrochen, und als das in den Jahren 1921-1922 immer häufiger geschieht, von blitzähnlichen Linien, die sich über die ganze Seite ziehen'. Georges Didi-Huberman, *Das Nachleben der Bilder: Kunstgeschichte und Phantomzeit nach Aby Warburg*, trans. by Michael Bischoff (Berlin: Suhrkamp, 2010), p. 413. Translation by author.

57 'war ihm doch bewußt, daß er ein Gleichnis seiner eigenen Situation lieferte, wenn er die Schlangen zwischen den Zähnen der Tänzer in Walpi zeigte'. Ibid., p. 402. Translation by author.

trans. by Massimo Zannini (Milano: Jaca Book, 2006).

Didi-Huberman, Georges, *Das Nachleben der Bilder: Kunstgeschichte und Phantomzeit nach Aby Warburg*, trans. by Michael Bischoff (Berlin: Suhrkamp, 2010).

Dietheuer, Franz, 'Das Programm der Apsisfresken in St. Jakob zu Kastelaz ob Tramin', *Der Schlern*, 65 (1991), 307-330.

Düriegl, Ursula, *Die Fabelwesen von St. Jakob in Kastelaz bei Tramin: Romanische Bilderwelt antiken und vorantiken Ursprungs* (Wien, Köln, Weimar: Böhlau, 2003).

Engels, Eve-Marie, 'Darwin/Darwinismus', *Lexikon der Mensch-Tier-Beziehungen*, ed. by Arianna Ferrari and Klaus Petrus (Bielefeld: transcript, 2015), 69-73.

Garber, Josef, *Die romanischen Wandgemälde Tirols* (Wien: Krystall, 1928).

Gombrich, Ernst H., *Aby Warburg: Eine intellektuelle Biographie* (Hamburg: Europäische Verlagsanstalt, 1981).

Luff, Robert, 'Bestiale est hominem nolle scire: Bildungsprogramm und Weltbild im deutschen 'Lucidarius' und auf den romanischen Fresken in St. Jakob/Kastelaz bei Tramin', *Der Schlern*, 74 (2000) 2, 99-119.

Martinek, Manuela, *Wie die Schlange zum Teufel wurde: Die Symbolik in der Paradiesgeschichte von der hebräischen Bibel bis zum Koran* (Wiesbaden: Harrassowitz, 1996).

Metzger, Wolfgang, 'Die Apsisfresken von St. Jakob in Kastelatz', *Der Schlern*, 44 (1970), 200-222.

Michaud, Philippe-Alain, *Aby Warburg and the Image in Motion*, trans. by Sophie Hawkes (New York: Zone Books, 2004).

Mundkur, Balaji, *The Cult of the Serpent: An Interdisciplinary Survey of Its Manifestations and Origins* (Albany: State University of New York, 1983).

Myss, Walter and Posh, Benedikt, *Die vorgotischen Fresken Tirols* (Wien: Herder, 1966).

Reichardt, Gerd, "De homine et portentis': Die Apsismalereien von St. Jakob bei Tramin', *Archäologie der Phantasie: Vom 'Imaginationsraum Südtirol' zur 'longue durée' einer 'Kultur der Phantasmen' und ihrer Wiederkehr in der Kunst der Gegenwart*, ed. by Elmar Locher, and Hans Jürgen Scheuer (Bozen: sturzflüge, 2012), S. 97-109.

Tetzlaff, Ingeborg, *Romanische Kapitelle in Frankreich: Löwe, Schlange, Sirene und Engel* (Köln: DuMont, 1976)

Theil, Edmund, *St. Jakob in Kastelaz* (Bozen: Athesia, 1978).

Warburg, Aby M., 'Der Eintritt des antikisierenden Idealstils in die Malerei der Frührenaissance', Id., *Die Erneuerung der heidnischen Antike: Kulturwissenschaftliche Beiträge zur Geschichte der europäischen Renaissance*, ed. by Horst Bredekamp and Michael Diers (Berlin: Akademie Verlag, 1998), 173-176.

Warburg, Aby, *Schlangenritual: Ein Reisebericht* (Berlin: Klaus Wagenbach, 1988).

Warburg, Aby, *Werke in einem Band*, ed. by Martin Treml, Sigrid Weigel, and

Perdita Ladwig (Berlin: Suhrkamp, 2010).

Warburg, Aby, and W. F. Mainland, 'A Lecture in Serpent Ritual', *Journal of the Warburg Institute*, vol. 2, no. 4 (Apr., 1939), 227-292. (www.jstor.org/stable/750040)

Wittkower, Rudolf, *Allegorie und der Wandel der Symbole in Antike und Renaissance* (Köln: DuMont, 1984).

FLAVIA PALMA

CHANGING SHAPES: HUMAN AND ANIMAL METAMORPHOSES IN STRAPAROLA

Giovan Francesco Straparola's *The Pleasant Nights* is one of the first instances of substantial employment of literary fairy tales in a collection of novellas,[1] which have been defined as an 'amalgama di materiali diversi',[2] including, in particular,

> unmistakable tales of magic, novelle in the manner of the *Decameron*, *exempla*, fables, twenty-three translations from the Latin *Novellae* of Girolamo Morlini, and two vernacular novelle (in the *bergamasco* and in the *pavano*).[3]

The marvellous and irrational elements that Straparola implanted in many tales, which probably influenced his decision to label them 'favole' instead of novellas,[4] may be a symbol of his conscious intention to break away from the Boccaccian model. Among the several examples of fairy tales pivoting on magic that we can find in *The Pleasant Nights*, a group of stories are particularly useful to prove Straparola's attitude towards the

---

1   This collection of tales is divided into two volumes: the first one was published for the first time in 1550, the second in 1553.

2   Donato Pirovano, 'Introduzione', in Giovan Francesco Straparola, *Le piacevoli notti*, ed. by Donato Pirovano (Roma: Salerno, 2000), pp. ix-l (p. xx).

3   Donato Pirovano, 'The literary fairy tale of Giovan Francesco Straparola', *The Romanic Review*, 99 (2008), 281-96 (p. 281). See also Renzo Bragantini, *Il riso sotto il velame. La novella cinquecentesca tra l'avventura e la norma* (Firenze: Olschki, 1987), p. 79; Daria Perocco, 'Trascrizione dell'oralità: il gioco delle forme in Straparola', in *Favole parabole istorie. Le forme della scrittura novellistica dal Medioevo al Rinascimento*, Atti del Convegno di Pisa, 26-28 ottobre 1998, ed. by Gabriella Albanese, Lucia Battaglia Ricci, and Rossella Bessi (Roma: Salerno, 2000), pp. 465-81 (pp. 475-76); Francisco Vaz da Silva, 'The Invention of Fairy Tales', *The Journal of American Folklore*, 123, 490 (2010), 398-425 (p. 419).

4   See Marga Cottino-Jones, *Il dir novellando: modello e deviazioni* (Roma: Salerno, 1994); Pirovano, *Introduzione*.

novella tradition: four 'favole' (II 1; III 3; III 4; VIII 4) are inhabited either by animals which can transform into human beings, or by men who can metamorphose into animals whenever they wish. Stith Thompson points out that 'although popular beliefs have [...] ascribed superhuman wisdom to some animals, there is even tendency for folk tradition to minimize the differences between man and beast. Sometimes heroes may assume either quality at will'.[5] The employment of such transformations, usual in the fairy tales, collides with the verisimilitude principle, typical of the novella form. While wit and eloquence were the quintessential weapons of the Boccaccian heroes, in *The Pleasant Nights* these same abilities seem inadequate. In order to triumph, many of Straparola's protagonists have to resort to magic or supernatural helpers.

This essay suggests that Straparola's divergence from the novella tradition (and its values) also passes through the representation of the relationship between human and animal nature. Indeed, the metamorphosis into different species, which is so natural and unconstrained in *The Pleasant Nights*, can be seen as a form of distrust in the power of human capabilities, whose exclusive employment seems ineffective. Straparola's attitude towards man's abilities may be read as a consequence of the precarious and unsafe socio-cultural climate of the mid-sixteenth century. In those days, many writers were sceptical of men's faculties to face adverse fortune and did not peaceably adhere to the novella tradition as founded in the fourteenth century, in which eloquence and cleverness were pivotal.

Since Boccaccio's *Decameron*, the novella has always been associated with plausibility, supplying a verisimilar representation of the events. Marga Cottino-Jones points out:

> From the *Decameron*, which provided the established model, on to the several collections of the XVth and early XVIth centuries, the novella had established itself as a short prose form articulated on a narrative discourse aimed at realistically representing the world of human experience under the assumption of a coherent view of 'reality', in terms of the mimetic, the visual, the physical, and the rational.[6]

5   Stith Thompson, *The Folktale* (New York: The Dryden, 1951), p. 245. It
     should be pointed out that Propp's theories on the fairy tales could be easily
     applied to several stories in *The Pleasant Nights*. See Vladimir Propp,
     Morfologia della fiaba (Torino: Einaudi, 1988).
6   Marga Cottino-Jones, 'Princesses, Kings, and the Fantastic: A Re-Vision of
     the language of Representation in the Renaissance', *Italian Quarterly*, 37,

The few sixteenth-century theoretical texts devoted to the genre also consider verisimilitude to be one of its fundamental aspects. In a section of his *Dialogo de' giuochi* (1572), Girolamo Bargagli introduces a discussion on the fantastic and the employment of magic in the *Decameron*,[7] concluding that

> men belle e meno perfette si tengono quelle [*scil.* novelle] che maghe, incanti e cose fatate contengono. E però lasciate cotali favole alle simplici fanciullette, qualcuna di caso verisimile ne narrarete, quando da comandamento di vegghia a ciò sarete astretti.[8]

In fact, Boccaccio allows for a scant employment of irrational and magic elements in his novellas, and generally uses them with satirical aim, having smart characters cheat on silly ones by taking advantage of their credulity. This is the case of donno Gianni (IX 10), or Bruno and Buffalmacco (VIII 3). Only very few novellas deal with magic seriously, as Bargagli himself points out with reference to the tales of messer Torello (X 9), madonna Dianora (X 5), and Nastagio degli Onesti (V 8), suggesting that the first two could be considered verisimilar in the fourteenth century since people still believed in necromancy, while the third is a sort of exception to the rule.[9]

In particular, only one novella in the *Decameron* pivots on the supposed transformation of a human being into an animal thanks to magic. It is the case of the aforementioned donno Gianni, who persuades a poor worker and his silly wife that he can transform women

---

143-146 (2000), 173-84 (p. 173).

7 On Bargagli's ideas about magic in the novellas see Cottino-Jones, *Princesses*. On the use of the term 'favola' in the novella tradition see also: Stefano Calabrese, *Gli arabeschi della fiaba. Dal Basile ai romantici* (Pisa: Pacini, 1984); Enrico Malato, 'La nascita della novella italiana: un'alternativa letteraria borghese alla tradizione cortese', in *La novella italiana*, Atti del Convegno di Caprarola, 19-24 settembre 1988, 2 vols (Roma: Salerno, 1989), I, 3-45; Hermann Wetzel, 'Premesse per una storia del genere della novella. La novella romanza dal '200 al '600', in *La novella italiana*, I, 265-91.

8 Girolamo Bargagli, *Dialogo de' giuochi*, in Nuccio Ordine, *Teoria della novella e teoria del riso nel Cinquecento* (Napoli: Liguori, 1996), pp. 137-155 (p. 145). ['Those [novellas] which contain magic, enchantments and fairy-tale topics are less beautiful and less perfect. Therefore, left these tales to simple girls, you will narrate a verisimilar case, if you are requested']. Unless otherwise stated, translations from Italian are mine.

9 See *Dialogo de' giuochi*, p. 144.

into horses. This proves to be nothing but a practical joke loaded with sexual innuendo, since the priest shows off his (fake) magical powers in order to have an intercourse with the worker's wife. By contrast, many examples of metamorphoses of men into animals, and vice versa, can be found in Straparola, without showing a humorous quality or aim. In 'favola' II 1 a young prince is born as a pig, but, despite any appearances, he speaks and thinks as a human being and, as the tale unfolds, he proves to be able to literally take off his animal skin, displaying a beautiful human body. Tale III 3 narrates the story of a princess, whose twin is a snake that transforms into a beautiful young girl, while in III 4 Fortunio is magically endowed with the power to transform into a wolf, an eagle, or an ant, as he likes. Finally, in VIII 4, Dionigi learns how to use necromantic arts and manages to change his shape, metamorphosing into animals or objects.

At first glance, Straparola seems to support the idea of man's superiority over every other living creature. Indeed, at the beginning of tale II 1, the narrator Isabella asserts:

> Quanto l'uomo, graziose donne, sia tenuto al suo creatore che egli uomo e non animale brutto l'abbia al mondo creato, non è lingua sì tersa né sì feconda, che in mille anni a sofficienza il potesse isprimere.[10]

In this 'favola' and in III 3 some expressions seem to strengthen the anthropocentric traditional Renaissance idea. In the former, the pig prince uses his animal hide as a sort of garment and transforms into a real man only after his pigskin has been destroyed. This pigskin is often referred to as 'la puzzolente e sporca pelle' (p. 103), 'la sua spoglia porcina' (Ibid.), 'la puzzolente scorza' (p. 104),[11] apparently suggesting that these animal features are to be seen as a degradation of human kind. The idea of the beastly shape as a mere cover which hides a superior nature has

---

10   Giovan Francesco Straparola, *Le piacevoli notti*, ed. by Donato Pirovano, 2 vols. (Roma: Salerno, 2000), p. 95. For all other references to this edition of *The Pleasant Nights* I will give only the corresponding page number after each quotation. The English translations of Straparola's passages are from: *The Nights of Straparola*, trans. by W. G. Waters, 2 vols. (London: Lawrence and Bullen, 1894). ['Fair ladies, if man were to spend a thousand years in rendering thanks to his Creator for having made him in the form of a human and not of a brute beast, he could not speak gratitude enough', *The Nights of Straparola*, 1, p. 58].

11   'The foul and dirty skin of the pig' (*The Nights of Straparola*, 1, p. 63); 'the dirty pig's hide' (Ibid.); 'his dirty hide' (*The Nights of Straparola*, 1, p. 64).

a correspondence in tale III 3, where the snake Samaritana plays the role of the magical helper in favour of her human sister, Biancabella. In particular, Biancabella's recovery, after she has had her eyes gouged and her hands cut off, begins when Samaritana takes on a human shape. As soon as she heals her sister, she also naturally lays down the 'squalida scorza di biscia' (p. 210) and reveals a beautiful female body.[12] However, the idea that the animal-like shape is just a temporary and debasing mask, as suggested by the use of the term 'squalida scorza' [dismal skin], may be challenged by Samaritana's own characterisation. On the one hand, Biancabella calls her 'sorella' [sister] while she is still a snake; on the other hand, her behaviour does not change, even after her transformation. This may be explained by the fact that the fairy tale does not make a strong distinction between animals and men, being often populated by talking animals behaving as human beings.

Nevertheless, one may wonder why Straparola, despite the narrator Isabella supports human superiority over animals, adopts a point of view close to the one of the fairy tale. In addition, several other elements in the four 'favole', on which this essay is focused, suggest the seminal function of the hero's metamorphosis into an animal in order to achieve a happy ending. These transformations are not degrading at all; on the contrary, they represent a weapon which guarantees the success that mere wit cannot ensure anymore, thus becoming a decisive strategy employed against the obstacles of Fate. Moreover, the figure of the hybrid is particularly relevant in these tales because it strengthens the challenge to the verisimilar novella tradition by highlighting the power of creatures which are neither totally humans nor totally animals, but blend at the same time aspects pertaining to these two different natures.

The 'favola' of the pig prince itself is meaningful from this point of view.[13] As many other fairy tales, it begins with a royal couple without heirs. One day the queen falls asleep in a garden and her beauty draws the attention of three fairies: the first two offer her the chance

---

12   'The scaly skin of the serpent' (*The Nights of Straparola*, 1, p. 135), but 'squalida' means 'dismal'.

13   The story of the pig prince has a long tradition, even though the protagonist may have different beastly shapes: he can be a bear, a serpent, a porcupine, etc. See Thompson, pp. 148-54; Suzanne Magnanini, *Fairy-Tale Science. Monstrous Generation in the Tales of Straparola and Basile* (Toronto-Buffalo-London: University of Toronto Press, 2008); Jan Ziolkowski, 'Straparola and the Fairy Tale: Between Literary and Oral Traditions', *The Journal of American Folklore*, 123, 490 (2010), 377-97.

to give birth to a beautiful baby endowed with every virtue; the third awkwardly affirms that the child will be 'tutto coperto di pelle di porco, e i gesti e le maniere che egli farà siano tutti di porco' (pp. 96-97).[14] Nobody knows about the fairies' gifts, not even the queen, who soon gives birth to a little pig, to everyone's astonishment. The horror that this unexpected offspring's features generate suggests that the prince will have a debasing reputation because of his bestial shape, so that his father, the king Galeotto, even ponders to kill the little monster. However, he finally acknowledges the creature as his own son:

> Ma pur rivolgendo nell'animo e discretamente pensando che 'l figliuolo, che si fusse, era generato da lui ed era il sangue suo, [...] volse al tutto non come bestia ma come animal razionale allevato e nodrito fusse. (p. 97)[15]

According to the third fairy's words, the young prince should effectively possess animal-like traits. However, by carefully considering the words the narrator employs in order to describe the little prince, it is possible to realise that the character is portrayed as a hybrid: on the one hand, he is introduced as the son ('figliuolo') of a man, having human parents and human blood; on the other hand, he displays the aspect of a non-human creature. His hybridity is strengthened by the fact that he is labelled not as a proper beast ('bestia'), but as a rational animal ('animal razionale'). The description of his behaviour, when he grows up, confirms that he blends animal shapes and (mostly) human attitude: indeed, he can 'umanamente parlare' (p. 97),[16] and manifests a strong will. In fact the only habit that characterises this 'porceletto' more as an animal than as a man is his rolling about in the mud, a practice which leads him to go back to the royal palace 'lordo e puzzolente' (Ibid.).[17] Nevertheless, he thinks as any other human being endowed with ingenuity and manages to persuade his parents, despite their horror, to find him a wife. They consent to their son's wish, but, when the

---

14   'In the skin of a pig, with a pig's ways and manners' (*The Nights of Straparola*, 1, p. 59).

15   'But when he debated in his mind and considered that this son, let him be what he might, was of his own begetting, [...] he made up his mind that the son should be brought up and nurtured like a rational being and not as a brute beast' (*The Nights of Straparola*, 1, p. 59)

16   'Talk like a human being' (*The Nights of Straparola*, 1, p. 60)

17   This phrase literally means 'dirty and smelly'.

chosen girl sees her dirty and beastly husband, she decides to kill him. However, the pig overhears her plans and counterattacks: 'Ma il porco fingendo di dormire con le acute sanne sí fortemente nel petto la ferí, che incontanente morta rimase' (p. 100).[18] The pig prince proves to be a skilful and manipulative creature. Indeed, he exploits his hybridity as a useful weapon: his animal appearance allows him to pass off his wife's death as a mere accident, while it is in fact a brutal assassination, so that his non-human features justify in a plausible way the results of his planned revenge.

The pig prince takes advantage of his parents' fear and love more than once: indeed, he manages to marry his dead spouse's sister, whom he kills too. At this point, he decides to get married with the third sister, Meldina, who proves to be truthfully fond of him. He tests her, as his previous wives, behaving as a real pig, but, realising that she is really nice as she seems, he decides to reveal to her his 'alto [...] segreto' (p. 103) [secret]:

> Sicurato adunque messer lo porco dalla moglie, si trasse la puzzolente e sporca pelle, e un vago e bellissimo giovane rimase, e tutta quella notte con la sua Meldina strettamente giacque. E impostole che al tutto dovesse tacere, perciò che era fra poco tempo per uscire di sí fatta miseria, si levò di letto, e presa la sua spoglia porcina, alle immondicie, sí come per lo adietro fatto aveva, si diede. (p. 103)[19]

The happy ending is close: Meldina gives birth to a beautiful baby and confesses to her mother-in-law that, when the prince is alone with her at night, 'si spoglia la puzzolente scorza' (p. 104),[20] disclosing a handsome human body. Thus, one night the royal parents destroy the pigskin, and Meldina and her prince inherit the realm, living happily ever after. Suzanne Magnanini considers the 'animal bridegroom tales', as this one, a disclosure of anxieties connected with sexuality and

---

18 Pretending to sleep, 'he struck her with his sharp hoofs and drove them into her breast so that he killed her' (*The Nights of Straparola*, 1, p. 61).

19 'Whereupon, being now sure of his wife's discretion and fidelity, he straightway shook off from his body the foul and dirty skin of the pig, and stood revealed as a handsome and well-shaped young man, and all that night rested closely folded in the arms of his beloved wife. But he charged her solemnly to keep silence about this wonder she had seen, for the time had not yet come for his complete delivery from this misery. So when he left the bed he donned the dirty pig's hide once more' (*The Nights of Straparola*, 1, p. 63)

20 'He casts off his dirty hide' (*The Nights of Straparola*, 1, p. 64).

wedding practices.[21] Nevertheless, she also suggests that this story may be linked to more specific worries sixteenth-century people felt with regard to monstrous births, which would lead to a disastrous crossing of the boundary between human and animal nature. She claims that 'before the reader arrives at the happy ending which reinstitutes the clear division between animals and humans that Isabella enthusiastically endorses at the beginning of the fairy tale, the monster must undergo a double erasure':[22] the first one corresponds to Galeotto's acknowledgement that the little pig is his own son and a rational creature; the second, begun after the third marriage, reaches its climax with the physical destruction of the pigskin.[23] From Magnanini's point of view, Straparola restores in this way the natural division between animals and human beings, which has been threatened by the events told in this 'favola' and specifically embodied by the pig prince himself.

Nevertheless, some details of the tale may also lead to another interpretation of this story. First of all, the protagonist tells Meldina that his capacity to change shape from pig to man is a secret he carefully hides, thus making the reader wonder whether he had previously transformed into man and for how long he has been able to do it. Giorgio Barberi Squarotti remarks that the prince himself is enslaved to dominating forces that determine his features, his virtues, and his behaviour.[24] However, this is not completely true, since he is perfectly able to control his transformations from animal into human, and vice versa, and he does not seem to be distressed by such condition. The only element that may suggest that this could be a painful status is the

---

21   Magnanini thinks that this fairy tale may also represent the threat embodied by the sexual desire of sixteenth-century young and unmarried Venetian aristocrats: '[...] the unmarried patrician pig boy's desire, frustrated in a restricted marriage market – albeit limited by his own deformity – becomes a disruptive force that challenges the throne and leads to acts of uxoricide. Whereas Galeotto erases the biological monster through scholastic theory and the physical destruction of the pig's hide, the social monster is erased through marriage to a woman who, although not of noble blood, succeeds in fulfilling her husband's desires and perpetuating the agnatic line'. This quotation is from: Suzanne Magnanini, 'Animal Anxieties: Straparola's *Il re porco*', in *The Italian Novella, A book of Essays*, ed. by Gloria Allaire (New York-London: Routledge, 2003), pp. 179-99 (p. 197).

22   *Fairy-Tale*, p. 110.

23   See *Fairy-Tale*, pp. 110-11.

24   Giorgio Barberi Squarotti, 'Problemi di tecnica narrativa cinquecentesca: lo Straparola', *Sigma*, 5 (1965), 84-108 (p. 94).

expression 'sì fatta miseria',[25] which alludes to his pig's shape. However, this comment may betray the narrator Isabella's ideological function,[26] stressing the idea of man's superiority over all other creatures. In fact, nothing in the habits of the prince implies that he may feel any sort of grief because of his hybridity. He keeps on putting on and off his animal hide, even after he has confessed his secret to Meldina, and continues rolling about in the mud, as he has always done. The prince's choice to remain a hybrid creature could be imputed to his desire to test his own family. Indeed, he seems to take advantage of his ability to change shape, and, if his parents had not destroyed his pigskin, he would have probably carried on this routine forever.

The character's unstable condition clearly favours him. Despite Isabella's opening statement about the supremacy of men over animals, the power that the protagonist exerts on people around him is due precisely to his hybridity which, being associated with a calculating wit, makes him a manipulative, but in the end, winning character. At the same time, the revelation of the existence of a pigskin which the character can put on and take off as he likes is meaningful: rather than an unquestionable hybrid experiencing both human and non-human nature, the pig prince may be considered as a pseudo-hybrid, namely a man hidden under an animal hide. As Lewis Seifert suggests, he may have always been 'a human in the skin of a pig'.[27] The end of the 'favola' can imply that the pigskin was in fact an instrument that the protagonist exploited and merely covered his humanity without really affecting it.

The pig prince is not the only character in Straparola's *The Pleasant Nights* who takes advantage of his animal features. In 'favola' III 4, Fortunio discovers that he has been adopted and decides to leave his home; his adoptive mother grieves his decision and curses him by foretelling that Fortunio will be eaten by a mermaid as soon as he goes at sea. During his journey, he meets an eagle, a wolf, and an ant: he solves a discussion

---

25 'This misery' (*The Nights of Straparola*, 1, p. 63).
26 For the ideological function of the narrator, see Gérard Genette, *Narrative Discourse. An Essay in Method* (Ithaca-New York: Cornell University Press, 1980).
27 Lewis C. Seifert, 'Pig or Prince? Murat, d'Aulnoy, and the Limits of Civilized Masculinity', in *High Anxiety. Masculinity in Crisis in Early Modern France*, ed. by Kathleen P. Long (Kirksville, Missouri: Truman State University Press, 2002), pp. 183-209 (p. 189). See also Lewis C. Seifert, 'Animal-Human Hybridity in d'Aulnoy's *Babiole* and *Prince Wild Boar*', *Marvels & Tales*, 25, 2 (2011), 244-260.

arisen between them, and the three thank him by endowing him with the ability to take on and off their features as he likes. He later falls in love with princess Doralice, and manages to enter her chamber by transforming into an eagle and an ant, and make her finally fall in love with him. After their wedding, Fortunio embarks on a voyage at sea and, as his mother had predicted, he is eaten by a siren, but he finally sets himself free by metamorphosing into an eagle. Before the end of the tale, he avenges on his adoptive mother by temporary turning into a wolf and eating her.

This short summary displays how the employment of magic and the protagonist's subsequent ability to become an animal are essential for the development of the events in this 'favola', in which the hero can triumph only by changing his shape at the right moment. Moreover, if compared with the pig prince, Fortunio is undoubtedly a hybrid creature, since the three animals he meets give him the chance to literally transform into different beasts. At the same time, if magic is a necessary weapon, which gives him the chance to conquer Doralice, escape from the mermaid, and take his revenge on his enemies, also the shrewdness in employing his magical powers is important.

The episode of his getaway from the siren is particularly meaningful, and deserves a further insight. Doralice waits more than two years before undertaking a voyage with her little son to the place where Fortunio was eaten by the mermaid. As soon as her ship arrives, she takes out a brass apple for her child to play with. The apple draws the attention of the siren, who asks for the object, promising in exchange to let Doralice see Fortunio up to his chest. As soon as the woman sees her husband, the siren goes away, bringing her prisoner along. At this point, the child starts playing with a silver apple and the same scene takes place, but this time the mermaid releases Fortunio up to his knees. In the end, in exchange of a golden apple, the siren makes her hostage completely visible, unadvisedly giving him the chance to transform into an eagle and fly away from her:

> Fortunio vedendosi fuori delle onde e sopra il dorso della sirena in libertà, tutto giolivo, senza interponere indugio alcuno, disse: 'Deh, fuss'io un'aquila!'; e questo detto subito aquila divenne, e levatosi a volo, sopra l'antenna della galea agevolmente salì e ivi, tutti i marinai vedendo, abbasso disceso, nella propria sua forma ritornò. (p. 230)[28]

---

28   'Now, as soon as Fortunio felt that he was quite clear of the water, and resting free upon the back of the Siren, he was filled with great joy in his heart, and,

The episode opposes two hybrid creatures: on the one hand, the siren, who is half fish and half woman; on the other hand, a man who is able to abandon his human shape and metamorphose into different animals. Interestingly enough, the first loses and the second wins because they are endowed with different human vices or qualities. Fortunio displays his wit by spotting and seizing the right moment to perform his metamorphosis, thus cunningly resorting to his magical powers. Nevertheless, his smartness and readiness would have been worthless if he could not have relied on magic, becoming an eagle. At the same time, he has the chance to show his shrewdness also thanks to the siren's greed, so that this semi-human sea creature is branded with one of the typical human vices recurring in both the novella and the fairy tale traditions.

Thus, in 'favola' III 4 Fortunio and the mermaid, as two hybrid beings, clearly represent a blend of human and non-human aspects, the former connected to their virtues or vices, the latter primarily represented by their (temporary or permanent) outward features.

'Favola' VIII 4 gives a further instance of the relationship between magic, wit, and metamorphosis, telling the story of the apprentice Dionigi and his master, the tailor (and secret necromancer) Lattanzio.[29] This tale is slightly different from the other ones which deal with the topic of metamorphosis in *The Pleasant Nights*, because its protagonist, who is described as 'diligente e accorto' (p. 553),[30] learns how to use magic without any external help, just spying Lattanzio while he practises necromancy at night. After having realised that he has been cheated, the master decides to avenge this 'theft' by preventing Dionigi to return to his human shape after metamorphosing into a horse. At this point, a long series of transformations into animals and objects, made by both

---

without hesitating for a moment, he cried out, 'Ah! would that I were an eagle', and scarcely had he ceased speaking when he was forthwith transformed into an eagle, and, having poised himself for flight, he flew high above the sail yards of the galley, from whence – all the shipmen looking on the while in wonder – he descended into the ship and returned to his proper shape' (*The Nights of Str>parola*, 1, p. 151)

29   For the oriental, in particular Indian, antecedent of this 'favola', see Giuseppe Rua, *Tra antiche fiabe e novelle. Le 'Piacevoli notti' di messer Gian Francesco Straparola* (Roma: Loescher, 1898), pp. 36-38. For other general indications about the sources of the tales in *The Pleasant Nights*, see Giuseppe Rua, 'Intorno alle *Piacevoli notti* dello Straparola', *Giornale storico della letteratura italiana*, 16 (1890), pp. 218-83.

30   'Industrious and prudent' (*The Nights of Straparola*, 2, p. 102)

the protagonist and the antagonist, takes place. While escaping from Lattanzio, Dionigi meets the daughter of the king, Violante, and makes her fall in love with him. The climax is reached in front of the king at the end of the tale when Dionigi metamorphoses into a pomegranate and the necromancer, transformed into a cock, eats all the fruit's grains but one. The grain turns into a fox and devours Lattanzio, who is still in the shape of a cock. Dionigi eventually resumes his human features, marries Violante and the two live happily ever after.

As in 'favola' III 4, magic gives Dionigi the chance to become a hybrid creature, but the character's magical abilities are even more powerful than those of his companions in the other tales, since he is able to transform not only into other living creatures, but also into objects, preserving his capability to think and feel as a sentient being. Indeed, while he is at the royal palace, Dionigi can spend all his time with Violante by becoming a ruby ring not to be discovered and returning man when possible. As happened with the animal metamorphosis, transforming into an object does not compromise the character's human nature:

> Partita dal padre Violante e andata nella sua camera e chiusa sola dentro, si mise a piagnere, e preso il robino, l'abbracciava, basciava e stringeva, maladicendo l'ora che il medico in queste parti era venuto. Vedendo il robino le calde lagrime che da i be' occhi giù scorrevano e i profondi sospiri che dal ben disposto cuore venivano, mosso a pietà si converse in umana forma […]. (p. 559)[31]

Violante keeps on treating him as a sentient being, so that she is said to be in love with the ruby itself ('tutta accesa dell'amor del robino', p. 559),[32] as if Dionigi-the-object and Dionigi-the-man were exactly the same. This is confirmed by the fact that Dionigi himself, although assuming the features of a ruby, still feels and thinks as a man, indeed he can both see ('vedendo') and be moved to piety ('mosso a pietà').

---

31   'When she had gone out of her father's presence, Violante went forthwith to her own chamber, and having fastened the door thereof, in her solitude she began to weep, and took the ruby and embraced and kissed it and pressed it to her heart, cursing the hour in which the physician had come across her path. As soon as the ruby saw the hot tears which fell from the lovely eyes of the princess, and heard the deep and woeful sighs which came from her loving heart, it was moved to pity, and straightway took upon itself the form of Dionigi, …' (*The Nights of Straparola*, 2, p. 108).

32   'Deeply enamoured of the ruby' (*The Nights of Straparola*, 2, p. 108).

Some scholars have criticised Straparola's characters by stressing their passivity or lack of reasoning due to the author's choice of the fairy tale. Marga Cottino-Jones, for example, asserts that his 'favole'

> seem to convey an ideology that questions the power of reason and of human enterprise, while opening for the individual confined to a condition of passivity and impotence, a way out through the irrational and the magic.[33]

Besides, Donato Pirovano points out that intelligence is not a characteristic of Straparola's protagonists, whose success is made possible just because external helpers assist them.[34] However, this is not a rule in *The Pleasant Nights*: in the tales concerning transformations from man into animal, and vice versa, even though the protagonists sometimes receive their supernatural ability from some external helpers, they clearly show a sort of awareness about when and how they should resort to their magical powers, thus exploiting their hybrid nature.[35]

Although he criticises the powerlessness of Straparola's characters' intelligence and initiative, Barberi Squarotti asserts that in his 'favole' 'l'astuzia non ha parte se non è aiutata dalla protezione soprannaturale' (p. 96),[36] which implies the presence of a certain 'amount' of wit. While these characters win thanks to their magical power, they also show to be able to employ their magical abilities in the right moment and way. This implies that, although wit is not the hero's most important virtue anymore, it is still present, but paired with supernatural abilities, which gradually gain momentum. There are several hints (in particular in tales III 3, III 4 e VIII 4) which may prove this idea, such as Fortunio's and Doralice's different behaviours in the episode of the getaway from the siren. Doralice knows both the magical powers of her husband and the mermaid's lust for the precious apples, however she does not take advantage of the situation, nor does she even realise that she could exploit the siren's covetousness in order to free her husband. She is passively satisfied by the chance she is given to see Fortunio. By contrast, Fortunio immediately takes

---

33 *Princesses*, p. 176. See also Barberi Squarotti, p. 95; Giancarlo Mazzacurati, *Forma e ideologia (Dante, Boccaccio, Straparola, Manzoni, Nievo, Verga, Svevo)* (Napoli: Liguori, 1974), p. 92.

34 *Introduzione*, p. xxvi.

35 See also Ruth Bottigheimer, 'Straparola's *Piacevoli Notti*: Rags-To-Riches Fairy Tales As Urban Creations', *Merveilles & contes*, 8, 2 (1994), pp. 281-96.

36 'Shrewdness has no effectiveness if it is not helped by supernatural protection'.

the opportunity to escape by resorting to his magical powers in order to regain his freedom: thus, he manifests a much more acute shrewdness. The same thing can be suggested for the pig prince: while his three wives either hate or love him without any specific reason, all his actions stem from a cunning plan aimed at conquering what he desires. He threatens his parents in order to get married; he pretends that the assassination of his first wife was involuntary; he exaggerates his 'animal-like' habits in order to try his wives. Furthermore, Dionigi is described as 'sì diligente e accorto, che quanto gli era dimostrato, tanto imparava' (p. 553).[37] This description could perfectly fit one of Boccaccio's heroes. The same fight between Dionigi and Lattanzio is played on the basis of a cunning employment of magic: they both decide to transform into an animal (or an object) according to the specific situation they are living in and the reactions of their respective enemy. Their conflict seems a battle of wits rather than a magic competition. As Marziano Guglielminetti suggests, the final scene of this 'favola' portrays a series of metamorphoses 'con una ricerca del gratuito e dell'assurdo non dettata da (barocca?) volontà di meraviglia, ma piuttosto dal calcolo intellettuale (rinascimentale?) delle possibilità che l'uomo ha per sfuggire alla propria sorte'.[38] Interestingly enough, Dionigi wins over Lattanzio taking up the features of an 'astuta e sagace volpe' (p. 561).[39]

Straparola does not deny the importance of wit, but exposes its inescapable fallibility if not helped by (extra-human) forces. Thus, the characters of the tales here analysed pass from human into animal shapes, or vice versa, in order to overcome the impediments they have to face. They maintain the acuteness and the ability to cunningly guess when to employ their talents, which is a typical feature of the novella tradition's characters. However, they finally reach their aims because they can resort to magic, leading to a voluntary, even if temporary, relinquishment of their human nature. Thus, Straparola's employment of fairy tales may reveal a sort of distrust in the values on which the novella tradition is based; men cannot use only their 'human' abilities, in particular their wit, in order to change their condition or to overcome the difficulties that Fate

---

37 'An industrious and prudent lad, who learnt with ease whatever his master attempted to teach him' (*The Nights of Straparola*, 2, p. 102)
38 Marziano Guglielminetti, *La cornice e il furto. Studi sulla novella del '500* (Bologna: Zanichelli, 1984), p. 35. ['Characterised by a research of absurdity which is due not to the (Baroque?) desire of marvel, but to the (Renaissance?) smart assessment of the chances that men have in order to be successful'].
39 'A crafty cunning fox' (*The Nights of Straparola*, 2, p. 109)

imposes on them. Thus, the hybrid protagonists of the tales here analysed embody the necessity to resort to something which goes beyond what pertains to human virtues and attitudes. The connection with the world of animals and magic is an instrument to stress the problematic condition of men living in an unstable world: human beings are not considered as powerful as they were in the previous decades, so that they need to resort to something which is non-human or extra-human in order to have some chance to succeed. Accordingly, as Giancarlo Mazzacurati observes, in Straparola the protagonist of the verisimilar (and Boccaccian) novella is contaminated with the fairy tales' main characters 'a sottolineare e testimoniare la crisi della civiltà borghese, tra Cinquecento e Seicento, preludio della rifeudalizzazione secentesca'.[40]

This clearly emerges in an episode recounted in 'favola' III 4, where Straparola resorts to the story narrated by Boccaccio in novella III 6 by rewriting and making it appropriate for the fairy tale climate of his narration. The apparent proximity between Fortunio's courtship of Doralice and Ricciardo's seduction of Catella in *Decameron* III 6, as observed by Giuseppe Rua and Donato Pirovano,[41] displays in fact two completely different realities. Although in both cases the young man is alone with his beloved and tries to persuade her to surrender to him, the authors' points of view are completely different. In Boccaccio, the skirmish between Catella and Ricciardo is offered in many details and structured on logical relations of cause and effect guaranteed by both the seminal role of reason and the persuasive effect of rational words. After Catella had slept with Ricciardo, believing he was her husband, the young suitor clearly exposes to her the likely options: she can either tell everybody the truth, thus risking being considered herself an adulteress, or keep silent. The narrator of the novella says: 'diede tanto luogo la ragione alle vere parole di Ricciardo, che ella cognobbe esser possibile a avvenire ciò che Ricciardo diceva'.[42] Although she decides not to reveal

---

40 Mazzacurati, p. 91. ['In order to highlight and testify to the sixteenth- and the seventeenth-centuries crisis of the middle-class culture, forerunner of seventeenth-century re-feudalisation']. See also Bottigheimer, p. 291; Cottino-Jones, *Il dir novellando*, pp. 143-44; Magnanini, *Animal Anxieties*, pp. 187-88. On the contrary, Pirovano thinks that Straparola, by choosing to employ the fairy tale in his collection, merely aimed to reach a wider reading public, interested in this kind of production (*The Literary Fairy Tale*, p. 283).

41 See Rua, *Tra antiche fiabe*; for Pirovano's statement, see Straparola, p. 223n.

42 See Giovanni Boccaccio, *Decameron,* ed. by Vittore Branca, in *Tutte le opere di Giovanni Boccaccio* (Milano: Mondadori, 1976), vol. 4, p. 288. ['She

what happened, she is still unfriendly towards Ricciardo. At this point, he displays once more his eloquence, persuading her with his 'dolcissime parole' [sweet words] to become his lover for good.[43] Boccaccio portrays here the power of speech, through which the hero can both assuage the woman's hostility, leading her to rationalise the situation and its consequences, and persuade her to do what he desires. The determining role of words is also shown by the space given to the characters' direct speeches. On the contrary, in Straparola's 'favola' the logical sequence of the events is less apparent, and Fortunio's talk, shortened to only a few lines, plays a less decisive role than Ricciardo's. Although Fortunio finally convinces Doralice to be kind to him, the author pays much more attention to Fortunio's cunning employment of his magical powers, which allows him to metamorphose several times into an eagle and an ant, thus entering the bedroom of the princess without being recognised or seen by anyone but her. While in the *Decameron* Ricciardo eventually conquers Catella by exploiting his wit, his initiative, and his rhetorical abilities, the relationship between Fortunio and Doralice always pivots on magic, which connects human beings to a world that goes beyond human' limits. Fortunio manages to approach Doralice just because he can metamorphose into different animals; he loses her because of a mythological creature, the siren; he goes back to his happy life thanks to his magical powers. As Ricciardo, Fortunio displays an evident initiative, but his success is clearly connected to the right employment of magic. Despite some outward similarities, Boccaccio's novella and Straparola's 'favola' are two different products, which derive from two different visions of the world: while in the former, the smart use of speech is enough in itself, in the latter, words have to be supported by the aid of supernatural forces and non-human abilities.

Thus, the employment, within a collection of novellas, of several fairy tales may imply Straparola's desire to manifest his distrust not in human intellect itself, but in its omnipotence, as canonised in the novella tradition. Herman Wetzel suggests that

> quando non si crede più che l'uomo sia capace di dominare la fortuna e di autocontrollarsi, gli autori cercano allora rifugio in quel

---

evaluated Ricciardo's true words by employing her reason, and realised that what he said could really happen'].

43   Boccaccio, p. 288.

genere letterario con il quale tradizionalmente si esprime l'utopia della soddisfazione dei desideri: la fiaba.[44]

Indeed, the fairy tale clearly testifies to the need for extra- and non-human forces in order to win over the obstacles created by Fate. By choosing to frequently employ stories of magic in his collections of tales, Straparola indirectly hints that the path of the verisimilar novella tradition as founded by Boccaccio is not possible anymore. This is also a consequence of the different socio-cultural climate in which Straparola wrote.

The mid-sixteenth century was a period full of uncertainties, which did not allow men to gain successful positions relying exclusively on their abilities. Ruth Bottigheimer, for example, suggests that Straparola's 'rags-to-riches' fairy tales should be considered as 'compensatory narratives' aimed to counterbalance sixteenth-century Venetians' economic difficulties: 'It must have seemed that only magic could accomplish reversals of fortune and social elevation' (p. 291). In general, the political, economic, and religious events undermined the certainties on which Italian culture had established itself for a long time. As a result of this difficult and uncertain historical climate, magic and irrationality entered literary production, including the novellas, which were usually devoted to a verisimilar representation of life. As Pirovano points out, Straparola voluntarily created a connection between his *Pleasant Nights* and the novella tradition through the employment of framing devices.[45] In this way the adjustments he made in his 'favole' within a canonised structure evidently foreground his scepticism towards the values of the traditional novella. Human abilities are inadequate on their own, since Fate can unexpectedly change every man's condition. With regard to the influence of the socio-cultural climate on Straparola's tales, Giorgio Barberi Squarotti writes that these stories are the symbol of a society which does not believe anymore 'all'iniziativa individuale, allo scatto della forza e dell'intelligenza che vale a conquistare il successo, a vincere l'ordine stabilito delle classi' (p. 92).[46]

The sixteenth-century precariousness is reflected into the fairy tale's instability due to the combined action of fate and magic. By resorting

---

44 Wetzel, p. 275. ['When people do not believe anymore that man is able to dominate Fate and to control himself, authors resort to the literary genre which traditionally expresses the utopia of the desires' satisfaction: the fairy-tale'].
45 *The Literary Fairy Tale*, p. 284.
46 'In personal initiative, in the employment of human strength and wit in order to be successful, to win the fixed social order'.

to tales belonging to this specific genre, Straparola could portray the necessary characteristics of the only possible hero of this new world, that is, a man who goes beyond human nature in order to survive and achieve his goals. Thus, the metamorphosis from man into animal, and vice versa, is not a mere consequence of the employment of the fairy tale, but can be read as an evident representation of the need to draw from other forces, which can guarantee success in a world where nothing is sure, nor permanent. The fairy tale allows for the representation of the hybridity of the new winning hero, who has to pair wit with extra- and non-human qualities. Interestingly enough, Dionigi, who is introduced as a smart and cunning boy, chooses to become a necromancer instead of a wealthy tailor, clearly explaining to his worried father that his decision has a practical reason. He wants to earn as much money as he can and magic is the best way to achieve this aim:

> Padre, [...] pregovi che non vi affanate, ancor ch'io non abbia apparato l'arte del sarto sì come era il desiderio vostro, percioché io n'apparai un'altra che ne sarà di maggior utile e contento. State adunque cheto, padre mio diletto, né vi smarrite, percioché presto vedrete il profitto ch'io fei, e del frutto la casa e la famiglia sovenir protrete. (p. 555)[47]

The metamorphosis into an animal is not a debasement of the human being, but a way to become successful: it is a new weapon, useful for a clever user.[48] From this point of view, hybridity embodies the necessity to be more than just a man. In Straparola, we are far from the Machiavellian image of the prince as the embodiment of both the fox and the lion: the Florentine writer exploited the animal metaphors in order to lively symbolise two different human qualities. On the contrary, in *The Pleasant Nights*, the chance to voluntarily become an animal implies a temporary relinquishment of human nature, and it is, at the same time, a

---

47 'My father, [...] I beg you that you will cease to disquiet yourself because I have not learnt the trade of a tailor as was your intention and desire, forasmuch as I have acquired the mastery of another art which will be of far greater service to us in the satisfying of our wants. Therefore, my dear father, do not disturb yourself or be sorrowful, because I will soon let you see what great profit I am able to make, and how, with the fruits of my art, you will be able to support your family and keep good cheer in your house' (*The Nights of Straparola*, 2, p. 104)

48 Giorgio Barberi Squarotti incidentally suggests that the metamorphosis into animals may represent the difficulties of changing one's own condition in a rigidly organised society (see p. 99).

successful way for humans in order to reach a happy ending. Isabella's statement at the beginning of 'favola' II 1, where she declares that men are superior to animals by God's will, is a perfect representation of the Renaissance standpoint, which is nevertheless challenged in the mid- and late sixteenth century, when confidence on the power of human abilities does not appear to be a viable option anymore. The only chance for a man to overcome the unpredictable forces which often act against him, at times with no apparent reason, is to become something Other than merely human, that is, a hybrid creature, who merges cleverness and magical powers to reach the non- and extra-human.

Thus, the four stories of metamorphosis here analysed can be considered as a meaningful representation of the growing distrust in the supremacy of the man's hallmark, that is wit, as the only winning weapon against Fate. The idea that cleverness alone is an inadequate tool and the growing importance given to supernatural forces imply the precarious condition of men in Straparola's times. The precariousness of the author's cultural and historical climate likely steered his literary choices: the employment of the fairy tale allows him to display his scepticism with regard to the traditional novella's hero. The protagonists of his tales of metamorphosis, although maintaining some traces of wit, see magic as a decisive instrument that may lead to success.

Bibliography

Barberi Squarotti, Giorgio, 'Problemi di tecnica narrativa cinquecentesca: lo Straparola', *Sigma*, 5 (1965), 84-108.
Bargagli, Girolamo, *Dialogo de' giuochi*, in Nuccio Ordine, *Teoria della novella e teoria del riso nel Cinquecento* (Napoli: Liguori, 1996), pp. 137-155.
Boccaccio, Giovanni, *Decameron*, ed. by Vittore Branca, in *Tutte le opere di Giovanni Boccaccio* (Milano: Mondadori, 1976), vol. 4.
Bottigheimer, Ruth, 'Straparola's *Piacevoli Notti*: Rags-To-Riches Fairy Tales As Urban Creations', *Merveilles & contes*, 8, 2 (1994), 281-96.
Bragantini, Renzo, *Il riso sotto il velame. La novella cinquecentesca tra l'avventura e la norma* (Firenze: Olschki, 1987).
Calabrese, Stefano, *Gli arabeschi della fiaba. Dal Basile ai romantici* (Pisa: Pacini, 1984).
Cottino-Jones, Marga, *Il dir novellando: modello e deviazioni* (Roma: Salerno, 1994).
Cottino-Jones, 'Princesses, Kings, and the Fantastic: A Re-Vision of the Language of Representation in the Renaissance', *Italian Quarterly*, 37, 143-146 (2000), 173-84.

Genette, Gérard, *Narrative Discourse. An Essay in Method* (Ithaca-New York: Cornell University Press, 1980).

Guglielminetti, Marziano, *La cornice e il furto. Studi sulla novella del '500* (Bologna: Zanichelli, 1984).

Magnanini, Suzanne, 'Animal Anxieties: Straparola's *Il re porco*', in *The Italian Novella, A book of Essays*, ed. by Gloria Allaire (New York-London: Routledge, 2003), pp. 179-99.

Magnanini, Suzanne, *Fairy-Tale Science. Monstrous Generation in the Tales of Straparola and Basile* (Toronto-Buffalo-London: University of Toronto, 2008).

Malato, Enrico, 'La nascita della novella italiana: un'alternativa letteraria borghese alla tradizione cortese', in *La novella italiana*, Atti del Convegno di Caprarola, 19-24 settembre 1988, 2 vols. (Roma: Salerno, 1989), I, pp. 3-45.

Mazzacurati, Giancarlo, *Forma e ideologia (Dante, Boccaccio, Straparola, Manzoni, Nievo, Verga, Svevo)* (Napoli: Liguori, 1974).

Perocco, Daria, 'Trascrizione dell'oralità: il gioco delle forme in Straparola', in *Favole parabole istorie. Le forme della scrittura novellistica dal Medioevo al Rinascimento*, Atti del Convegno di Pisa, 26-28 ottobre 1998, ed. by Gabriella Albanese, Lucia Battaglia Ricci, and Rossella Bessi (Roma: Salerno, 2000), pp. 465-81.

Pirovano, Donato, 'Introduzione', in Giovan Francesco Straparola, *Le piacevoli notti*, ed. by Donato Pirovano (Roma: Salerno, 2000), pp. ix-l.

Pirovano, Donato, 'The Literary Fairy Tale of Giovan Francesco Straparola', *The Romanic Review*, 99 (2008), 281-96.

Propp, Vladimir, *Morfologia della fiaba* (Torino: Einaudi, 1988).

Rua Giuseppe, 'Intorno alle *Piacevoli notti* dello Straparola', *Giornale storico della letteratura italiana*, 16 (1890), 218-83.

Rua Giuseppe, *Tra antiche fiabe e novelle. Le 'Piacevoli notti' di messer Gian Francesco Straparola* (Roma: Loescher, 1898).

Seifert, Lewis C., 'Pig or Prince? Murat, d'Aulnoy, and the Limits of Civilized Masculinity', in *High Anxiety. Masculinity in Crisis in Early Modern France*, ed. by Kathleen P. Long (Kirksville, Missouri: Truman State University Press, 2002), pp. 183-209.

Seifert, Lewis C., 'Animal-Human Hybridity in d'Aulnoy's *Babiole* and *Prince Wild Boar*', *Marvels & Tales*, 25, 2 (2011), 244-60.

Straparola, Giovan Francesco, *Le piacevoli notti*, ed. by Donato Pirovano, 2 vols. (Roma: Salerno, 2000).

The Nights of Straparola, trans. by W. G. Waters, 2 vols. (London: Lawrence and Bullen, 1894).

Thompson, Stith, *The Folktale* (New York: The Dryden, 1951).

Vaz da Silva, Francisco, 'The Invention of Fairy Tales', *The Journal of American Folklore*, 123, 490 (2010), 398-425.

Wetzel, Hermann, 'Premesse per una storia del genere della novella. La novella romanza dal '200 al '600', in *La novella italiana*, Atti del Convegno di Caprarola, 19-24 settembre 1988, 2 vols. (Roma: Salerno, 1989), I, pp. 265-91.

Ziolkowski, Jan, 'Straparola and the Fairy Tale: Between Literary and Oral Traditions', *The Journal of American Folklore*, 123, 490 (2010), pp. 377-97.

Simone Rebora

'IT'S AS SEMPER AS OXHOUSEHUMPER!' JAMES JOYCE'S ANIMALIZATION OF THE HUMAN

Animals are definitely not a central element in James Joyce's fiction. As famously stated by the author himself, its final goal was that of building an 'epic of the human body',[1] where every single component of human history, culture, and language could find a perfect correspondence in a corporeal (human) function. Animals were thus marginal in this overarching design, as confirmed by their quasi-total absence in some of the most famous representations of Joyce's novels, such as the 'Linati Schema' for *Ulysses* (traced by the author in 1920, two years before the publication of the novel),[2] or the 'Diagram of Finnegans Wake' by László Moholy-Nagy,[3] drawn eight years after the publication of Joyce's last work.

The most famous among Joyce's animals is probably the cat, starting from the one that gives the title to the short story *The Cat and the Devil*, conceived by Joyce in 1936 as an ironic fable for his grandson Stephen and published posthumously in 1964. Notwithstanding its centrality, however, the cat has a rather passive role in the story, being used by the

---

1   Derek Attridge (ed.), *James Joyce's Ulysses: A Casebook* (New York: Oxford University Press, 2004), p. 261.

2   Animals appear only with a reference to 'Zoology' in the symbolic dimension of the episode 'Circe'. Joyce drew another schema in 1921, known as 'Gilbert schema': here the Horse is the only animalist symbol which appears, in relation to the episode 'Nestor'. Both schemas were entirely published by Richard Ellmann, *Ulysses on the Liffey* (London: Faber and Faber, 1972), plates not numbered.

3   See László Moholy-Nagy, *Vision in Motion* (Chicago: Theobald Publishers, 1947), p. 347. Moholy-Nagy's diagram represents synthetically the narrative and symbolic structure of *Finnegans Wake* by overlapping a table (that lists the protagonists and shows their functions on several conceptual levels) with a circular pattern (where multiple time dimensions are crossed by places and events, like sun rays propagating from a central medallion which bears the initials of the author). No traces of animals – except for the mythical phoenix – can be found in this complex diagram.

protagonist as a simple device to deceive the devil.[4] Another cat appears
at the beginning of *Ulysses*, just after the presentation of its protagonist
Leopold Bloom. Here the cat seems to play a more active role, by driving
the thoughts of the protagonist and even 'talking' to him:

> – Milk for the pussens, he said.
> – Mrkgnao! the cat cried.
> They call them stupid. They understand what we say better than we
> understand them. She understands all she wants to. Vindictive too. Cruel.
> Her nature. Curious mice never squeal. Seem to like it. Wonder what I
> look like to her. Height of a tower? No, she can jump me. (*U* 4.24-29)[5]

It has been noted that, through these few observations, Leopold
Bloom (and Joyce with him) shows a new, even revolutionary attitude
towards animals, trying to assume their point of view. However, David

---

4   The story is re-elaborated by Joyce from the French folk tradition. The
     protagonist is the lord major of Beaugency, who makes a pact with the devil
     for building a bridge that connects the two parts of the city, separated by a
     wide river. The price for the work is the first soul which will cross the bridge.
     To deceive the devil, the lord major carries out a plan: on the day of the
     inauguration, he shows up on the bridge holding a cat and a bucket of water.
     'The lord major put[s] the cat down on the bridge and, quick as thought,
     splash! he emptie[s] the whole bucket of water over it', James Joyce, *The Cat
     and the Devil* (London: Faber and Faber, 1965), pages not numbered. What
     follows can be easily inferred, while Joyce ironically focuses on the angry
     reaction of the devil, who 'speaks a language of his own called Bellsybabble
     which he makes up himself as he goes along but when he is very angry he can
     speak quite bad French very well', *The Cat and the Devil*, page not numbered.
     Another short story focused on cats has been more recently published, once
     again based on a letter to the author's grandson. However, in *The Cats of
     Copenhagen*, cats do not even appear. Joyce laments their absence in the city,
     proceeds with piquant – and slightly surreal – observations on Danish people,
     and ends up with a proposal: 'When I come to Copenhagen again I will bring
     a cat and show the Danes how it can cross the road without any instructions
     from a policeman', James Joyce, *The Cats of Copenhagen* (New York:
     Scribner, 2012), page not numbered.
5   References to Joyce's classic works follow the conventions shared by Joycean
     scholars: for *A Portrait of the Artist as a Young Man* and *Ulysses*, an indication
     of the chapter/episode followed by line numbers, see James Joyce, *A Portrait
     of the Artist as a Young Man*, ed. by Richard Ellmann (New York: Viking
     Press, 1964); James Joyce, *Ulysses*, ed. by Hans Walter Gabler, et al. (New
     York and London: Garland Publishing, 1984); for *Finnegans Wake*, an
     indication of the page followed by line numbers, see James Joyce, *Finnegans
     Wake* (New York: Viking Press, 1939).

Rando suggests carefulness in interpreting Bloom's thoughts. In fact, 'Bloom misinterprets the cat's desires because he projects his feelings onto her'.[6] Specifically, he projects onto the cat his feelings towards his wife, Molly (who is waiting in the bed not for him, but for Blazes Boylan), thus making the incidental allusion to cruelty sound more like a sexist remark, than a testimony of veterinary wisdom.

This example helps to understand how animals, while confined to the margins of Joyce's fiction, actually play a decisive role in it. When confronting themselves with human beings, they can show all the limits and contradictions of human nature, even through a simple reflection. Their potential is exponentially increased when fully integrated in the complex symbolic network that structures a work like *Ulysses*. Animals inhabit almost all of its eighteen episodes: from the threatening dogs in 'Proteus' and 'Cyclops', to the lugubrious rat in 'Hades' and the stuffed owl in 'Ithaca', they all seem to play some symbolic function in the novel, allowing Robert Haas to build an alternative 'Linati schema', where body parts are fully substituted by animals.[7]

All this symbolic proliferation considered, two episodes in *Ulysses* show animals which particularly dominate the scene. In 'Circe', the starting point is offered by the Homeric subtext (with Ulysses' companions transformed into pigs), but what happens to the protagonist is even more complex and disturbing. Through the episode, dominated by a hallucinatory style, Bloom is turned into a woman ('Dr Bloom is bisexually abnormal', *U* 15.1775-1776), a sow, and then a bull, while suffering the most degrading verbal and physical vexations ('Lynch him! Roast him! He's as bad as Parnell was', *U* 15.2898-2899). On the one hand, this process of continuous transformation seems to finally break the boundaries between human and animal, in a sort of automated *danse macabre* that also reveals the mechanical nature of life itself. As noted by Maud Ellmann, in 'Circe' 'human superiority is exposed as a delusion, based on the repression of the animal and the mechanical,

---

6   David Rando, 'The Cat's Meow: *Ulysses*, Animals, and the Veterinary Gaze', *James Joyce Quarterly*, 46.3-4 (2009), 529-543 (p. 536).

7   See Robert Haas, 'A James Joyce Bestiary', *ANQ*, 27.1 (2014), 31-39 (p. 32). The schema shows 'the major animal symbol for each episode and the pages on which it occurs. Each animal acts either in sympathy or in opposition to Stephen or Bloom [...], and the symbols then function in such key structural roles as character painting or development, or conflict or kinship, at the end even extending to the reader too', Haas, pp. 31-39 (p. 31).

repressions that return with a vengeance'.[8] However, this degradation also entails a much more refined overthrow when read through all its symbolic significance. As suggested by Maureen O'Connor:

> The irony of the humiliating 'degradation' of Bloom being rendered female and animal is clear here. Through these transformations, Bloom is finally identified with the most elevated of sacrificial victims, Jesus Christ and, Joyce's personal political hero, Charles Stewart Parnell.[9]

It is not by chance that, in the first episode of *Ulysses*, Ireland is given the epithet of 'Silk of the kine' (*U* 1.403), recalling a xenophobic, centuries-old tradition passed down by the colonial authority, that 'rendered Ireland both 'feminine' and animal-like'.[10] Surprisingly enough, Joyce does not contradict this tradition, but he brings it to the core of an even more complex literary structure.

The image of the bull, in fact, stands out also in another episode of *Ulysses*, 'Oxen of the Sun'. Here the animal symbol goes as far as to give the title to the whole chapter, by connecting once again animality with sexual alteration (being the ox a castrated bull). The episode is one of the most complex in the novel, because it combines at least thirty-one different literary styles, ranging from early incantations to future forms of communication.[11] As suggested by Susan Bazargan, there are three central themes in 'Oxen of the Sun', that is maternity, history and

---

8   Maud Ellmann, '*Ulysses*: Changing into an Animal', *Field Day Review*, 2 (2006), 75-93 (p. 75).

9   Maureen O'Connor, '"Mrkgnao!' Signifying Animals in the Fiction of James Joyce', in *James Joyce. Critical Insights*, ed. by Albert Wachtel (Ipswich: Salem Press, 2013), pp. 101-119 (p. 117).

10   O'Connor, pp. 101-119 (p. 104).

11   According to Cheryl Temple Herr, 'the fundamental events that occur in 'Oxen of the Sun' are straightforward enough. Reluctant to return home, Bloom goes to the Holles Street hospital to see whether Mina Purefoy, who has been in labor for three days, has given birth. There he joins a group of medical students who drink with abandon while discussing questions of birth, disease, theology, and death'. But the composition of the episode is far from straightforward: 'The reader begins to see – and this is crucial to understanding 'Oxen of the Sun' – the rigorously controlled clock-time of Bloomsday breaking apart and overlapping so that competing nighttime chronologies seemingly distort the go-ahead daytime narrative', Cheryl Temple Herr, 'Difficulty: 'Oxen of the Sun' and 'Circe'', in *The Cambridge Companion to 'Ulysses'*, ed. by Sean Latham (New York: Cambridge University Press, 2014), pp. 154-170 (pp. 157-158).

language, 'intertwined in the chapter by virtue of a substructure of a womb through which we travel 'historically' on a plane of language constantly evoking and transforming the past'.[12] This complex construction seems to follow an evolutionary path, but ends up in a final disintegration of linguistic conventions: after imitations of Latin historiographers (*U* 14.7-32), Elizabethan prose writers (*U* 14.277-333), and authors like John Ruskin (*U* 14.1379-1390) and Thomas Carlyle (*U* 14.1391-1439), the chapter is closed by what Joyce defined 'a frightful jumble of Pidgin English, nigger English, Cockney, Irish, Bowery slang and broken doggerel'.[13] An apparent triumph of chaos over language,[14] that suggests anyway the possibility of communication beyond its orderly limits. If we follow the interpretation of Laura Pelaschiar, in fact, the episode 'ends not so much in chaos […] as in difference, which somehow bypasses the idea of one-ness and unity or uniformity'.[15] However, to better understand how this widespread difference is connected with human language and animality, a further step must be taken towards Joyce's last work, famously defined as his 'book of the dark'.[16]

In the fifth chapter of the first book of *Finnegans Wake*, an entire passage is dedicated to the description of a mysterious letter, discovered

---

12   Susan Bazargan, 'Oxen of the Sun: Maternity, Language, and History', *James Joyce Quarterly*, 22.3 (1985), 271-280 (pp. 271-272).

13   James Joyce, *Letters*, ed. by Stuart Gilbert (New York: Viking Press, 1966), I, p. 139.

14   As noted by Margot Norris, 'Joyce's model for the episode, that the styles of literary history recapitulate the embryonic development of a child in the womb, may have been deliberately misleading. The developing styles do not improve wisdom, insight, or significance and do not issue in the miracle of perfect modern literary language. Instead, the archaic perceptions and cognitive systems reveal the distortions produced by Possible Words, by the systems of beliefs, values, knowledge, and desire inscribed in language and expression and inherited from literary genres', Margot Norris, 'The (Im) possible Worlds of Joyce's 'Oxen of the Sun' Episode in *Ulysses*', *Genre*, 41.1-2 (2008), 95-123 (p. 121).

15   Laura Pelaschiar, 'Joyce the Museologist: 'Oxen of the Sun' as Joyce's Museum of Language', in *The Exhibit in the Text: The Museological Practices of Literature*, ed. by Caroline Patey and Laura Scuriatti (Oxford: Peter Lang, 2009), pp. 127-141 (p. 135).

16   See John Bishop, *Joyce's Book of the Dark, Finnegans Wake* (Madison: University of Wisconsin Press, 1986).

by the hen Biddy Doran on a pile of dung.[17] After three pages simply listing 'many names at disjointed times' (*FW* 104.5) through which this 'untitled mamafesta' (*FW* 104.4) had been known, the analysis seems to focus on its contents. However, what emerges from the description is the very impossibility of expressing the contents of the letter in a stable and unique way. The prose itself shatters into a series of cryptic allusions, thus making impossible a univocal interpretation of the passage:

> The proteiform graph itself is a polyhedron of scripture. There was a time when naif alphabetters would have written it down the tracing of a purely deliquescent recidivist, possibly ambidextrous, snubnosed probably and presenting a strangely profound rainbowl in his (or her) occiput. To the hardily curiosing entomophilust then it has shown a very sexmosaic of nymphosis in which the eternal chimerahunter Oriolopos, now frond of sugars, then lief of saults, the sensory crowd in his belly coupled with an eye for the goods trooth bewilderblissed by their night effluvia with guns like drums and fondlers like forceps persequestellates his vanessas from flore to flore. Somehows this sounds like the purest kidooleyoon wherein our madernacerution of lour lore is rich. [...] In fact, under the closed eyes of the inspectors the traits featuring the chiaroscuro coalesce, their contrarieties eliminated, in one stable somebody similarly as by the providential warring of heartshaker with housebreaker and of dramdrinker against freethinker our social something bowls along bumpily, experiencing a jolting series of prearranged disappointments, down the long lane of (it's as semper as oxhousehumper!) generations, more generations and still more generations. (*FW* 107.8-35)

These few lines give a significant example of the language in which *Finnegans Wake* was composed. The syntactic structure is that of the English language, fully recognizable through the use of articles, conjunctions, and pronouns, but also discernible through the apparent misprinting of numerous adjectives, verbs, and nouns. However, at a closer inspection, the text appears traversed by numerous interferences with many other languages, which create such a complex network of connections that the possible meanings of each single line could

---

17   Note that the letter is easily interpretable as a synecdoche of the entire book. According to Tindall, it is a 'concentrat[e] of the *Wake* itself', William York Tindall, *A Reader's Guide to Finnegans Wake* (Syracuse: Syracuse University Press, 1996), p. 12.

be extended exponentially. Roland McHugh devoted his work to the retrieval of all these sub-references in the text (and many other similar projects can now be found online).[18] In the passage reported here, he recognized a strong presence of Aramaic words in the central part (*'Ar* kidout'iun: science [...] *Ar* madénakrout'iun: literature [...] *Ar* lore: news'),[19] he confirmed the entomological and sexual subtext through some punctual remarks (*'Vanessa*, a genus of butterflies [...] Some moths emit sexual attractant odours at night'),[20] and he disentangled the last compound word, 'oxhousehumper' ('Hebrew letters: Aleph means 'ox', Beth means 'house', Gimel means 'camel'').[21]

Coming back to the interpretation of the passage, one of its possible meanings becomes a reflection on the timelessness of language, that could be interpreted through the most basic physiological functions (and also the references to digestion abound),[22] but that cannot be reduced to a stable form of knowledge (with the insisted allusions to proteiformity, coalescence, and continuous regeneration). However, towards the end, there comes the ironical revelation: the interpretation of the text could develop throughout eternity, while the final wisdom is as simple – or: as eternal – as ABC: it is already known from the beginning. And it is precisely here that the image of the ox finds once again its centrality, placing an animal (on which humans have exerted the harshest form of dominion) at the outset of language, knowledge, and time.

This ineluctable circularity also recalls *Finnegans Wake*'s cyclical structure, that derives directly from the Viconian conception of history

---

18   See for example *FinnegansWiki*, http://www.finnegansweb.com/wiki [accessed 28 August 2016]; *FinWake*, http://www.finwake.com [accessed 5 March 2017]; and *FWEET*, http://www.fweet.org [accessed 5 March 2017].

19   Roland McHugh, *Annotations to Finnegans Wake*, 4th edn (Baltimore: John Hopkins University Press, 2016), p. 107.

20   Ibid.

21   Ibid.

22   See for example the description of the 'hardily curiosing entomophilust' (*FW* 107.12-13), with 'the sensory crowd in his belly coupled with an eye for the goods trooth' (*FW* 107.15-16), an expression that also includes an allusion to teeth, from which the process of digestion begins. Incidentally note that the act of writing is later described as a distribution of excrements over the surface of the human body: 'he shall produce nichthemerically from his unheavenly body a no uncertain quantity of obscene matter not protected by copriright in the United Stars of Ourania [...] this Esuan Menschavik and the first till last alshemist wrote over every square inch of the only foolscap available, his own body' (*FW* 185.28-36).

– but with some relevant distinctions. As famously asserted by Joyce himself, Vico's philosophy was simply the basis for a much more extended construction: 'I don't take Vico's speculations literally. I use his cycles as a trellis'.[23] Arthur Walton Litz suggests giving less importance to these obvious debts, looking instead at deeper affinities, such as their shared interpretation of language, that Joyce brings to an even further stage:

> Vico looked upon language as fossilized history and sought to recover the past from the radical meaning of words; Joyce reversed this process and sought to create new verbal units which would embody the entire history.[24]

Through this reversing of perspectives, the dynamic remains the same, in the acknowledgment that etymology does not simply offer us an uncorrupted image of the past, but it may also be used as a creative vehicle to incite much more extended forms of knowledge.

This conception of language points towards another possible philosophical referent, Ludwig Wittgenstein. As noted by Megan Quigley,[25] in fact, the research paths of the Irish writer and the Austrian philosopher show a striking similarity, evolving from strictly logical conceptions of language towards more complex constructions.[26] Even though extremely different in their form and conception, with their mature productions both Joyce and Wittgenstein 'share a similar goal:

---

23   Mary and Padraic Colum, *Our Friend James Joyce* (London: Gollancz, 1959), p. 123.

24   Arthur Walton Litz, 'Vico and Joyce', in *James Joyce. Critical Assessments of Major Writers*, ed. Colin Milton (London-New York: Routledge, 2011), IV, pp. 287-297 (p. 295).

25   Megan Quigley, *Modernist Fiction and Vagueness. Philosophy, Form, and Language* (New York: Cambridge University Press, 2015), pp. 103-105.

26   A more rigid conception of language is supported in the *Tractatus Logico-Philosophicus* by Wittgenstein and the 'Study of Languages' by Joyce, conceived respectively in 1918 and 1899. The philosophy of language expressed in later works like *The Philosophical Investigations* and *Finnegans Wake* is, instead, deeply different. See Ludwig Wittgenstein, *Tractatus Logico-Philosophicus* (London: Routledge & Kegan Paul, 1922); James Joyce, *Occasional, Critical, and Political Writing*, ed. by Kevin Barry (Oxford: Oxford University Press, 2000), pp. 12-16; Ludwig Wittgenstein, *Philosophical Investigations*, 3rd edn, ed. by G.E.M. Anscornbe (New York: Macmillian, 1958).

'to run against the boundaries of language".[27] These are boundaries that entrap philosophical thinking into a pre-defined structure – as well as literary creation inside a sterilizing tradition – and that can be escaped only by talking what Wittgenstein called 'nonsense': 'not because it must be ridiculed, but because it is beyond the realm of logical analysis'.[28] As famously stated by the philosopher: 'in the end when one is doing philosophy one gets to the point where one would like just to emit an inarticulate sound'.[29] An inarticulate affirmation of life that Thomas Singer immediately connects with the 'yes' of Molly Bloom at the end of *Ulysses*,[30] but that can also be connected with the over-articulation of language in *Finnegans Wake*, when what remains in the end is a simple re-affirmation of reality. Another, even deeper similarity between Joyce and Wittgenstein, is in the recognition that 'the ordinary is the extraordinary, that the wonder of this world is not hidden behind any veil but is open to our view',[31] in its most common and trivial forms.

A confirmation for this idea can be found in Joyce's first novel, *A Portrait of the Artist as a Young Man* (originally published in 1916), where the image of the bull/ox/cow appeared for the first time, already charged with all its symbolic significance. The novel begins, in fact, with a fable narrated to the young Stephen Dedalus by his father:

> Once upon a time and a very good time it was there was a moocow coming down along the road and this moocow that was coming down along the road met a nicens little boy named baby tuckoo ...
> His father told him that story: his father looked at him through a glass: he had a hairy face.
> He was baby tuckoo. (*P* 1.1-7)

It has been noted how this passage already testifies to the decisive importance of language in Joyce's fiction. Stephen's identity is not determined by his auto-definition through language and thought but, in a sharp reversing of the Cartesian motto, it is externally defined by language itself.[32] However, the presence of the 'moocow' here is not

---

27 Quigley, p. 108.
28 Quigley, p. 137.
29 Wittgenstein, *Philosophical Investigations*, p. 261.
30 See Thomas C. Singer, 'Riddles, Silence, and Wonder. Joyce and Wittgenstein Encountering the Limits of Language', *ELH*, 57.2 (1990), 459-484 (p. 482).
31 Singer, pp. 459-484 (p. 462).
32 Singer, pp. 459-484 (p. 469).

purely casual. While Stephen tries in every way to structure his world into an ideal and harmonic order provided by language, what he meets in the course of the narrative is actually a disturbing reality, as shown by the successive appearance of cows, this time in a filthy cow-yard:

> But when autumn came the cows were driven home from the grass: and the first sight of the filthy cowyard at Stradbrook with its foul green puddles and clots of liquid dung and steaming bran troughs, sickened Stephen's heart. (*P* 2.128-132)

On this level, the *Portrait* becomes the story of a philosophical failure, a reversed *Bildungsroman* where the protagonist vainly tries 'to extricate an authentic 'self' from the mire of the lustful, sensual body, the 'inhuman' animal envelope of the soul',[33] while involuntarily showing the inconsistency of this ideal goal.

This reflection on language is recalled and deepened in *Finnegans Wake*, starting from a passage where animals are once again the protagonists. 'The Mookse and The Gripes' (*FW* 152.15-159.18) appears as a satirical adaptation of the well-known Aesopian fable *The Fox and the Grapes*, but also contains an overt parody of the *Portrait*'s incipit: 'Eins within a space and a wearywide space it wast ere wohned a Mookse' (*FW* 152.18-19). The Aesopian fox is thus transformed into a semi-anthropomorphic figure, the Mookse, that seems to occupy a very similar position to Stephen's in the *Portrait*. In fact, the Mookse engages in an exhausting debate with the Gripes met 'on the yonder bank of the stream that would be a river, parched on a limb of the olum' (*FW* 153.9-10). Their argument revolves around the dominance of space over time (or of time over space) and the two creatures seem in fact to rely on just one of the two dimensions: the Mookse lives in space, while the Gripes live in time. The Mookse (like Stephen) relies on a pre-ordered pattern of the world, while the Gripes (like Leopold Bloom in *Ulysses*) are more freely immersed in the flow of time. The discussion soon takes on parodic overtones (with insults such as: '–Unuchorn! –Ungulant! –Uvuloid! –Uskybeak!', *FW* 157.3-6) and ends up with the appearance of 'Nuvoletta', high in the sky, who decries the uselessness of the discussion. Here the tone comes to be more heartfelt, while the two prostrated disputant are taken away by two mysterious figures, and Nuvoletta dissolves herself into the stream, transforming it into a river:

---

33   O'Connor, pp. 101-119 (p. 107).

Then Nuvoletta [...] climbed over the bannistars; she gave a childy cloudy cry: *Nuée! Nuée!* She was gone. And into the river that had been a stream [...] there fell a tear, a singult tear, the loveliest of all tears. (*FW* 159.6-13)

This short story is part of the more extended narrative structure of *Finnegans Wake*, which can be described – at the risk of oversimplification – as the story of a family composed of a married couple (Humphrey Chimpden Earwicker and Anna Livia Plurabelle, generally known as H.C.E. and A.L.P.) and their three children (Shaun, Shem, and Issy). Evidently, the three protagonists of 'The Mookse and the Gripes' are personifications of the three children, while the story thematizes the competition between the two brothers, Shem and Shaun: narrated by Shaun, it is presented as the decisive argument for the affirmation of space over time (as evidenced by the linguistic effort to exclude any chronological reference), but it clearly fails its aim.[34]

Another passage from *Finnegans Wake* with a very similar structure is 'The Ondt and the Gracehoper' (*FW* 414.18-419.10), once again a parodic adaptation of an Aesopian fable, *The Ant and the Grasshopper*. It is interesting to notice how, in this passage, Joyce seems to follow more closely the Aesopian – or, more properly, the Lafontainian – model,[35] with the fall into disgrace of the grasshopper and the cynical imposition of the ant. Some slight inconsistencies in the characters, however, suggest a more complex interpretation. Sam Slote has noted how, for example, the Ondt's success appears to be contradictory:[36] more than in the accumulation of capital, he looks attracted by the possibility of interests; more than in procreation, he is interested in sexual satisfaction. With money investments and erotic achievements (accomplished with the lovers stolen from the Gracehoper), 'Joyce's Ondt betrays his parsimonious and usurious tendencies by gloating

---

34　Note for example the incipit of the passage: 'Eins within a space and a wearywide space it wast ere wohned a Mookse' (FW 152.18-19). The classic 'Once upon a time' is transformed into 'One within a space', and the German interferences help hiding the past tense of the verb 'to be': 'wohnen' is the German 'to reside', and the influence of German pronunciation might also transform the ancient English 'wast' (was) into 'vast'. However, this recall to ancient English introduces also a slight inconsistency in Shaun's operation, because 'there' is truncated into 'ere', ancient version of the preposition 'before'.

35　See Tindall, p. 229.

36　Sam Slote, 'The Prolific and the Devouring in 'The Ondt and the Gracehoper'", *Joyce Studies Annual*, 11 (2000): 49-65 (pp. 60-62).

over his reproductive faculties'.[37] In sum, by referring once again to the opposition between time and space, the space-living Ondt experiences his success in a time projection, thus invading an area where he has no actual expertise. A contradiction clearly understood by the Gracehoper, who closes the episode with a corrosive song of forgiveness:

> *Your feats end enormous, your volumes immense,*
> *(May the Graces I hoped for sing your Ondtship song sense!),*
> *Your genus its worldwide, your spacest sublime!*
> *But, Holy Saltmartin, why can't you beat time? (FW 419.5-8)*

This conclusion is interpreted by Slote 'as an indication of the mutually incomplete nature of the Ondt and the Gracehoper: each needs the other to exist'.[38] A relationship that reflects once again the intrinsic connection between the two conflicting brothers Shaun and Shem. Their conflict is continuously repeated throughout the book by means of different incarnations (like Mutt and Jute in *FW* 16.10-18.16, Butt and Taff in *FW* 338.5-355.7, and many others), always playing the same roles: Shem is the artist while Shaun is the politician, Shem lives in time while Shaun lives in space, Shem is related to the lower part of the (male) body, while Shaun is related to the upper part of the (male) body.

Referring to the inconclusive resolution of 'The Mookse and the Gripes', Edmund Epstein notes how this 'is not the ending that Shaun intended, but it is the one that he eventually has to produce, since his brother is the source of 'root language''.[39] At the same time, this basic form of language mastered by Shem cannot participate in the process of communication without coming into conflict with its more elevated counterpart. This situation highlights the double nature of language already represented by the double reading approach prompted by a text like *Finnegans Wake*: to be read partly on the surface and partly in depth, partly in the daylight and partly at night.[40] In a passage like 'The

---

37 Slote, pp. 49-65 (p. 62).
38 Slote, pp. 49-65 (p. 55).
39 Edmund Lloyd Epstein, *A Guide through Finnegans Wake* (Gainesville: University Press of Florida, 2009), p. 78.
40 As suggested by Sebastian Knowles, '*Finnegans Wake* (and this will be deeply controversial) must be read sober, in the cold light of day, which we did in class time, checking on plot elements, looking up references in the guidebooks, puzzling out the meaning of the text. But it must also (and this will not be controversial at all) be read at night, with the ear rather than the

Ondt and the Gracehoper', this duplicity is evidenced once again by McHugh's annotations, that help discerning the sub-language speaking beneath the surface of the text: the dominant semantic areas in these five pages are once again those of entomology and sexuality,[41] the same that were present in the passage about the 'untitled mamafesta' (*FW* 104.4).

To bring back this discourse to its historiographical context, it is necessary to notice that during the modernist decades, according to Philip Armstrong, the Romantic 'good savage' was finally domesticated, while 'it was time for art to break loose, go feral and return to a revitalizing savagery'.[42] In reaction to the horrors brought about by technological evolution and sustained by the development of psychanalysis, 'the negative therio-primitivism [...] was inverted to produce a redemptive therio-primitivism. Animality, at its most wild and untamed, was not the enemy of humanity, but its possible, perhaps its only, salvation'.[43] And *Finnegans Wake* seems to bring to the highest levels this re-affirmation of primitivism and animality, right through its peculiar interpretation of sexuality and entomology.

With regard to sexuality, it has been noted for example how 'genitalia, the instinctual centers of the being that are the power of Shem, are also the source of [...] deep, creative language'.[44] Shaun is uneasily aware of this and, as he makes efforts to become a writer, a donkey[45] maliciously asks him questions about this subject. But Joyce's scholars

---

eye, preferably over a glass of something, to bring the sounds of the text to wash over us as music, as a soundscape, as a flow of language, more id than ego, more nightmare than waking', Sebastian Knowles, *'Finnegans Wake* for Dummies', *James Joyce Quarterly*, 46.1 (2008), 97-111 (pp. 102-103).

41   Among the references related to entomology: 'jigger: flea *Tunga penetrons* [...] *L* (artif.) vespatilla: little wasp [...] *Gr* melissa: bee [...] *F* fourmi: ant [...] *G* Spinner: silkworm [...] elytra: hardened wing cases of beetles [...] *Heb* deborah: bee [...] *F* taon: gadfly [...] *Saturnia*: a moth genus [...] *Periplaneta*, a genus of cockroaches [...] *Myrmica*: a genus of ants [...] *Satyr*: a genus of butterflies [...] *Anopheles*, a genus of mosquitoes', McHugh, pp. 414-415; regarding sexuality: '*Da* hor: adultery [...] *Sl* ladybirds: lewd women [...] L libido: desire [...] *Arch* buss: kiss [...] Mozart: *Così Fan Tutte* (means 'this is how all women do it') [...] wittol: a conniving cuckold [...] *St* jade: whore [...] formication: sensation of ants fornication', McHugh, pp. 416-417.

42   Philip Armstrong, *What Animals Mean in the Fiction of Modernity* (London and New York: Routledge, 2008), p. 134.

43   Armstrong, p. 143.

44   Epstein, p. 77.

45   This is another animal transmutation of Shem, but also a personification of Shaun's genitalia, see Epstein, p. 22. Epstein also suggests that the Gripes are

have also tended to further exalt this aspect by fostering interpretations of *Finnegans Wake* that reduce its cosmological and overarching structure to the narrowed space of human genitalia (through a symbolic disposition of its five protagonists):

> To be more explicit, Joyce's five characters can be regarded as representing members of the propagating family, namely the penis and the testicles on the male side, and the labia of the vulva on the female side. Indeed, I am convinced that it would be proper to say that the universe of *Finnegans Wake*, which is, from one point of view, as boundless as infinity, could also be reduced, from another point of view, to the area immediately surrounding and encompassing the human genitals.[46]

However, Joyce's interpretation of sexuality is far from a pornographic exhibition. Indeed, it has been suggested that sexual intercourses are absent from the novel. Following once again Epstein's interpretation, the only one that actually occurs is located outside the book, in the hiatus that connects its interrupted final lines ('Till thousendsthee. Lps. The keys to. Given! A way a lone a last a loved a long the', *FW* 628.14-16) with the suspended incipit ('riverrun, past Eve and Adam's, from swerve of shore to bend of bay, brings us by a commodius vicus of recirculation back to Howth Castle and Environs', *FW* 1.1-3). While the stream of prose is fluidly recomposed here, that of the river appears contradictory, because in the final page the same river is moving towards the sea, while in the first page it is moving backwards, bound to Dublin. The intercourse is thus in the tidal movement of the river Liffey (symbolic transfiguration of the female protagonist A.L.P.): 'Every day, twice a day, the great act of love, the embrace and withdrawal of Anna Livia and the ocean, acts out an eternal love affair'.[47] And this embrace is clearly more significant than sex, because it also implies a form of liquefaction and fusion of the opposites, the kind of union that Julia Kristeva directly connects with the peculiar quality of Joyce's style:

---

nothing 'but the genitals of the Mookse, hanging 'bolt downright' from the branch of an elm', Epstein, p. 78.

46   Margaret Solomon, *Eternal Geomater. The Sexual Universe of Finnegans Wake* (Carbondale-Edwardsville: Southern Illinois University Press, 1969), p. 60.

47   Epstein, p. 13.

Liquidation, liquefaction, of both feminine and masculine, in the flow of a style, halting at no one identity – whether personal, ideological, or sexual – but knowing them all.[48]

Entomology is immediately linked to sexuality,[49] but it also enacts a different strategy for the achievement of this fusion of the opposites, namely through an affirmation of difference. In his influential reflections on animals, in fact, Jacques Derrida underlines the importance of speaking of them using the plural form:

Among nonhumans and separate from nonhumans there is an immense multiplicity of other living things that cannot in any way be homogenized, except by means of violence and willful ignorance, within the category of what is called the animal or animality in general.[50]

Fusion of the opposites, as a consequence, cannot be obtained through a sterile homogenization which, for Derrida, is not only 'a sin against rigorous thinking [...]; it is also a crime'.[51] This con-fusion exists instead in abundance and distinctions, and the irreducible proliferation of swarming insects in *Finnegans Wake* seems to accomplish this specific task,[52] as well as the proliferation of languages in its subtext.[53]

---

48 Julia Kristeva, 'Joyce 'The Gracehoper' or the Return of Orpheus', trans. by Louise Burchill, rev. by Jacques Aubert and Shari Benstock, in *James Joyce: The Augmented Ninth*, ed. by Bernard Benstock (Syracuse: Syracuse University Press, 1988), pp. 167-180 (p. 179-180).

49 One of the most significant slips of the tongue in 'The Ondt and the Gracehoper' is that between 'insect' and 'incest' right at the beginning: 'The Gracehoper [...] was always making ungraceful overtures [...] to commence insects with him' (*FW* 414.22-27).

50 Jacques Derrida, 'The Animal that Therefore I Am (More to Follow)', trans. by David Wills, *Critical Inquiry*, 28.2 (2002), 369-418 (p. 416).

51 Ibidem.

52 In addition, note that the name of its protagonist (Humphrey Chimpden Earwicker) synthesizes the wholeness of humankind (through his nickname 'Here Comes Everybody', *FW* 32.18-19), but also the encounter between animal sexuality ('humper' alludes to the camel, but also to dorsal sexual intercourse) and entomology (through the reference to the minute earwigs).

53 Another *pastiche* on which Joyce likes to indulge is that between entomology and etymology: 'Like the Gracehoper, Joyce displays in this fable 'his good smetterling of entymology' (417.4), a smattering, like German butterflies, of insect science grounded in interest in words and word origins', Margot Norris, 'The Animals of Finnegans Wake', *MFS: Modern Fiction Studies*, 60.3 (2014), 527-543 (p. 540).

According to Margot Norris, the function of animals 'is therefore part of the *Wake*'s project of setting human cultural figures and narratives into the much larger horizon of the spiritual world, on the one hand, and the natural world, on the other'.[54] Through the quasi-encyclopedic inclusion of entomological science, this horizon is expanded towards its extreme limits, generating an illusion of completeness that only few 'encyclopedic novels'[55] have reached in the history of literature.

The result, in its extreme linguistic and symbolic complexity, can be experienced as a form of noise: the white noise of overcharged communication systems or, more properly, the green noise of ecological difference.[56] That is, the 'background noise' of humans and animals living on the planet earth: the noise of life itself.

Bibliography

Armstrong, Philip. 2008. *What Animals Mean in the Fiction of Modernity* (London and New York: Routledge)

Attridge, Derek (ed.). 2004. *James Joyce's Ulysses: A Casebook* (New York: Oxford University Press)

Bazargan, Susan. 1985. 'Oxen of the Sun: Maternity, Language, and History', *James Joyce Quarterly*, 22.3: 271-280

Bishop, John. 1986. *Joyce's Book of the Dark, Finnegans Wake* (Madison: University of Wisconsin Press)

Colum, Mary and Padraic. 1959. *Our Friend James Joyce* (London: Gollancz)

Derrida, Jacques. 2002. 'The Animal that Therefore I Am (More to Follow)', trans. by David Wills, *Critical Inquiry*, 28.2: 369-418

Ellmann, Maud. 2006. '*Ulysses*: Changing into an Animal', *Field Day Review*, 2: 75-93

Ellmann, Richard. 1972. *Ulysses on the Liffey* (London: Faber and Faber)

Epstein, Edmund Lloyd. 2009. *A Guide through Finnegans Wake* (Gainesville: University Press of Florida)

---

54   Norris, pp. 527-543 (p. 528).
55   See Luc Herman and Petrus van Ewijk, 'Gravity's Encyclopedia Revisited: The Illusion of a Totalizing System in *Gravity's Rainbow*', *English Studies*, 90.2 (2009), 167-179.
56   In signal processing, white noise is a signal that has equal intensity on its entire frequency spectrum. Green noise is defined as 'the midfrequency component of white noise', Daniel L. Lau, Gonzalo R., Arce, and Neal C. Gallagher, 'Green-Noise Digital Halftoning', *IEEE*, 86.12 (1998), 2424-2444 (p. 2424), but is also known as 'the 'background noise' of the world, hence its name', Daniel J. Schneck, 'Music, the Body in Time, and Self-Similarity Concepts', *Journal of Biomusical Engineering*, 1 (2011), 1-9 (p. 8).

Haas, Robert. 2014. 'A James Joyce Bestiary', *ANQ*, 27.1: 31-39

Herman, Luc, and van Ewijk, Petrus. 2009. 'Gravity's Encyclopedia Revisited: The Illusion of a Totalizing System in *Gravity's Rainbow'*, *English Studies*, 90.2: 167-179

Herr, Cheryl Temple. 2014. 'Difficulty: 'Oxen of the Sun' and 'Circe'', in *The Cambridge Companion to 'Ulysses'*, ed. by Sean Latham (New York: Cambridge University Press)

Joyce, James. 1939. *Finnegans Wake* (New York: Viking Press)

– 1964. *A Portrait of the Artist as a Young Man*, ed. by Richard Ellmann (New York: Viking Press)

– 1965. *The Cat and the Devil* (London: Faber and Faber)

– 1966. *Letters*, ed. by Stuart Gilbert (New York: Viking Press), I

– 1984. *Ulysses*, ed. by Hans Walter Gabler, et al. (New York and London: Garland Publishing)

– 2000. *Occasional, Critical, and Political Writing*, ed. by Kevin Barry (Oxford: Oxford University Press)

– 2012. *The Cats of Copenhagen* (New York: Scribner)

Knowles, Sebastian. 2008. '*Finnegans Wake* for Dummies', *James Joyce Quarterly*, 46.1: 97-111

Kristeva, Julia. 1988. 'Joyce 'The Gracehoper' or the Return of Orpheus', trans. by Louise Burchill, rev. by Jacques Aubert and Shari Benstock, in *James Joyce: The Augmented Ninth*, ed. by Bernard Benstock (Syracuse: Syracuse University Press), pp. 167-180

Lau, Daniel L., Arce, Gonzalo R., and Gallagher, Neal C. 1998. 'Green-Noise Digital Halftoning', *IEEE*, 86.12: 2424-2444

McHugh, Roland. 2016. *Annotations to Finnegans Wake*, 4th edn (Baltimore: John Hopkins University Press)

Moholy-Nagy, László. 1947. *Vision in Motion* (Chicago: Theobald Publishers)

Norris, Margot. 2008. 'The (Im)possible Worlds of Joyce's 'Oxen of the Sun' Episode in *Ulysses'*, *Genre*, 41.1-2: 95-123

Norris, Margot, 2014. 'The Animals of Finnegans Wake', *MFS: Modern Fiction Studies*. 60.3: 527-543

O'Connor, Maureen. 2013. "Mrkgnao!' Signifying Animals in the Fiction of James Joyce', in *James Joyce. Critical Insights*, ed. by Albert Wachtel (Ipswich: Salem Press), pp. 101-119

Pelaschiar, Laura. 2009. 'Joyce the Museologist: 'Oxen of the Sun' as Joyce's Museum of Language', in *The Exhibit in the Text: The Museological Practices of Literature*, ed. by Caroline Patey and Laura Scuriatti (Oxford: Peter Lang), pp. 127-141

Quigley, Megan. 2015. *Modernist Fiction and Vagueness. Philosophy, Form, and Language* (New York: Cambridge University Press)

Rando, David. 2009. 'The Cat's Meow: *Ulysses*, Animals, and the Veterinary Gaze', *James Joyce Quarterly*, 46.3-4: 529-543

Schneck, Daniel J. 2011. 'Music, the Body in Time, and Self-Similarity Concepts', *Journal of Biomusical Engineering*, 1: 1-9

Singer, Thomas C. 1990. 'Riddles, Silence, and Wonder. Joyce and Wittgenstein Encountering the Limits of Language', *ELH*, 57.2: 459-484

Slote, Sam. 2000. 'The Prolific and the Devouring in 'The Ondt and the Gracehoper'', *Joyce Studies Annual*, 11: 49-65

Solomon, Margaret. 1969. *Eternal Geomater. The Sexual Universe of Finnegans Wake* (Carbondale-Edwardsville: Southern Illinois University Press)

Tindall, William York. 1996. *A Reader's Guide to Finnegans Wake* (Syracuse: Syracuse University Press)

Walton Litz, Arthur 2011. 'Vico and Joyce', in *James Joyce. Critical Assessments of Major Writers*, ed. Colin Milton (London-New York: Routledge), IV, pp. 287-297

Wittgenstein, Ludwig. 1922. *Tractatus Logico-Philosophicus* (London: Routledge & Kegan Paul)

Wittgenstein, Ludwig. 1958. *Philosophical Investigations*, 3rd ed., ed. by G.E.M. Anscornbe (New York: Macmillian)

Eirini Apanomeritaki

THE ANIMAL AS WRITER: CREATIVITY AND MELANCHOLIA IN MARIE DARRIEUSSECQ'S *PIG TALES*

'Bristles (the shame of it! but I will tell)
Began to sprout; I could no longer speak;
My words were grunts, I grovelled to the ground.'

(Ovid, XIV. 276-278)[1]

'The other is not only my enemy, his beastly jouissance is inside me: I am that beast, O animal, my soulmate, my brother!'

(Julia Kristeva, *Colette*)[2]

The question of the animal(s) and the implications of our co-existence with the animal world have occupied a central position in twentieth-century continental philosophy, with the influence of scholars such as Gilles Deleuze, Félix Guattari, and Giorgio Agamben still permeating contemporary animal narratives. At the same time, fantastic stories of animal-human metamorphoses have long enriched our imagination and challenged conventional ideas about other species, primarily through narratives about hybrid beings.[3] While animal studies have adopted less anthropocentric approaches to address contemporary anxieties in our relationship to animals, such as the extinction of species in capitalist societies, certain features which distinguish us from non-human creatures have rarely been negotiated. Language, for instance, and much more creativity, have always remained strictly human privileges,

---

1   Ovid, *Metamorphoses*, trans. by A. D. Melville, rev. edn. (Oxford: Oxford University Press, 2008), p. 333
2   Julia Kristeva, *Colette*, trans. by Jane Marie Todd (New York: Columbia University Press, 2004), p. 85.
3   Marina Warner explores the subjectivities of hybrid and metamorphosed beings in her book *Fantastic Metamorphoses, Other Worlds: Ways of Telling the Self* (Oxford: Oxford University Press, 2002).

as Agamben reminds us in the *Open: Man and Animal* (2002).[4] Towards the end of the previous century, Marie Darrieussecq's postmodern tale about a woman's transformation into a pig and a writer addressed the ongoing debate on human and animal interaction by focusing on the issue of interspecies. A narrative told by a pig's point of view is known to us from both Homer and Ovid; Darrieussecq's story though shifts our attention to a marginalized hybrid subject whose language, unlike its classic predecessors, can be now deciphered.

My enquiry into Marie Darrieussecq's debut novel *Pig Tales: A Novel of Lust and Transformation* (1996) focuses on the complex connections among creativity, grief and the trope of metamorphosis of the female subject as they are relayed from my interpretation of Darrieussecq's text.[5] I particularly focus on the emerging writer in combination with the emerging animal of the text exploring the '*écriture* de cochon,' [6] mastered by the pig after her nightmarish experiences and the death of her wolf mate. The schema of loss and mourning as triggering the act of writing, which has been theorised by Julia Kristeva in her work on depression and melancholia, assists my reading of Darrieussecq's story. Starting from a reading of Kristeva's *Black Sun*, I wish to suggest that it is possible to read the animal-human subject as melancholic and abject by considering that Pig's ability to write is prompted by a loss.[7] Once these links have been established, and taking my cue from Hélène Cixous, I suggest that a literary space can be created for hybrid identities, but not exclusively female ones.[8] I explore these two key thinkers because their writings resonate in Darrieussecq's story and particularly in her reimagining of female subjectivity and creative writing.[9] My approach in the discussion of a marginalised animal-

---

4   Giorgio Agamben, *The Open: Man and Animal*, trans. by Kevin Attell (Stanford: Stanford University Press, 2004), p. 36.

5   Marie Darrieussecq, *Pig Tales: A Novel of Lust and Transformation*, trans.by Linda Coverdale (London: Faber and Faber, 1997).

6   Marie Darrieussecq, *Truismes* (Paris: P.O.L, 1996), p. 10.

7   Julia Kristeva, *Black Sun: Depression and Melancholia*, trans. by Leon S. Roudiez (New York: Columbia University Press, 1989).

8   A number of scholars have emphasised Darrieussecq's hesitation to adopt a strictly political-feminist way of writing, especially the '*écriture* feminine' by Cixous. See Damlé (2014) and Jordan (2004).

9   In an illuminating discussion of *Pig Tales*, Shirley Jordan notes that Darrieussecq's writing can be read in light of Kristeva's theory of abjection, in *Contemporary French women's writing: women's visions, women's voices, women's lives, Vol. 37*, (New York: Peter Lang, 2004), p. 107.

human subjectivity is also informed by Giorgio Agamben's 'Homo Sacer', but also by the 'obscenely sexualized counterpart *mulier sacra*,' a concept suggested by Andrew Asibong in his article on Darrieussecq and Agamben.[10] Agamben's discussion of 'bare life' and exiled hybrid creatures from society allows a political reading of the pig's metamorphoses in contemporary Paris. Throughout the twentieth century, ideas on female creativity and animal rights further developed, so that stories of metamorphosis of female animals can be meaningfully read in the light of the above theorists.

A Pig on the Margins

Partly because *Pig Tales* alerts the reader on a number of contemporary issues, such as women's rights, animal exploitation and animal killing, totalitarianism, and, most critically the restrains on women's freedom, Darrieussecq's sharply satirical novel has attracted a significant amount of critical attention from animal and feminist studies, but also from fantastic, magical realist scholarship.[11] Scholars have either criticised the novel's grotesque portrayal of the pig woman as diminishing for female agency,[12] while others have praised the text's ecological concerns and the engagement with femininity which empowers alternative identities.[13] These alternative identities revisit existing, still relevant, debates on female creativity and identity. The concept of becoming-animal, theorized by Deleuze and Guattari, may shed light on the exploration of the constant shift of the protagonist between woman and animal, as Amaleena Damlé has argued.[14] Yet, the process of becoming does not consider female writers or creative animals. Damlé acknowledges the limitations of the Deleuzian model and adds

---

10   Andrew Asibong, 'Mulier Sacra: Marie Chauvet, Marie Darrieussecq and the Sexual Metamorphoses of 'Bare Life'', *French Cultural Studies*, 14.2, (2003), 169-177 (p. 171).

11   For example, Amaleena Damlé (2014) approaches Darrieussecq's novel in the light of Deleuze and Guattari's concept of 'becoming-animal' while Sallie Muirden (2008) emphasises the magical, allegorical aspect of the metamorphosis.

12   See Shirley Jordan (2004).

13   See Sanja Bahun (2011) and Anat Pick (2011).

14   Amaleena Damlé, *The Becoming of the Body: Contemporary Women's Writing in French* (Edinburgh: Edinburgh University Press, 2014), pp. 123-153.

that, what for Deleuze and Guattari is not possible, for Darrieussecq is imperative: 'the possibility of attaining political agency' is part of the metamorphosis she aims to narrate.[15] Indeed, by manipulating the trope of metamorphosis, an entirely unrealistic process, the story of an in-between human communicates empowering fantasies of hybrid selves which challenge the traditional concepts of the self and engage with progressive, even for some twentieth-first century audiences, formulations of female identity.

While many critics and readers have suggested that Franz Kafka's stories and, in particular, his 'Metamorphosis' (1915)[16] have inspired Marie Darrieussecq's *Pig Tales*, the writer herself has contested the comparison. In one of her interviews, Darrieussecq has addressed this issue explicitly, by explaining what lies at the heart of her interest in metamorphosis: 'everyone talked about Kafka, which is daft. It's completely different from Kafka. There's a gulf between Kafka and women, and Kafka and women's bodies.'[17] Darrieussecq's feminist treatment of metamorphosis focuses on women and the representations of the feminine body; issues which are absent in Kafka. However, both these fictional metamorphoses narrate socially constructed changes as the subjects metamorphose based on society's projections on them and their vulnerable position. Darrieussecq's choice of animal, one of exceptionally pink flesh, is based on the author's views of the society's treatment of women, like Kafka's choice of an insect for Gregor is evocative of his 'vermin-like' existence as an insignificant part of the early twentieth-century capitalist machinery.[18] Ten years after the publication of *Pig Tales*, Darrieussecq said that she was still interrogated on the choice of the animal in her best-selling novel. In the introduction to her 2006 novel, *Zoo*, she reiterates the reasons for this selection:

15   Damlé, p. 133.
16   Franz Kafka, 'The Metamorphosis,' trans. by Susan Bernofsky, intr. by David Cronenberg, 1st ed. (New York: W. W. Norton, 2014).
17   Marie Darrieussecq, interview to Fiona Cox in London, (2012) trans. by Fiona Cox. *Practitioners' Voices in Classical Reception Studies, Special Issue 2013: Contemporary Women Writers*. < http://www.open.ac.uk/arts/research/pvcrs/2013/darrieussecq> [accessed 5 March 2017]
18   In Darrieussecq's birthplace, Bayonne, there is a Ham Fair (the Foire au Jambon) held annually during the Easter weekend. The choice of pig may have been partly inspired from this festival.

we treat women as sow, more often than mare, cow, monkey, viper, tigress; more often than still as giraffe, leech, slug, octopus, or tarantula; and far more often than as a centipede, female rhino, or koala. It's simple.[19]

Here Darrieussecq specifies her own post-modern enquiry in *Pig Tales*: social projections on women affect their appearance and, their actions, but also, and perhaps primarily, their very identity. Indeed, as a third-wave feminist writer, from *Pig Tales* she has repeatedly addressed current affairs in French society through her writing, fictional or not, including sexual and political scandals, far-right policies, and most recently attacks on creativity and satire.[20]

Drawing upon a bleak vision of Paris where hybrid beings and all types of minorities of French society need to hide from the authorities and animal meat is sold in the black market, Pig has to hide in locations outside the borders of the Parisian community: the city sewers, an asylum, a forest, and an abandoned farm. The exclusion from society begins when the protagonist fails to comply with the standards of beauty and sex industry she is part of, due to her gradual change into a human sized sow. Finding it impossible to conceal her pink flesh and grey patches with makeup products, eventually her whole body succumbs to the pig nature. Hence, her male acquaintances, including her partner Honoré, progressively abandon her, as she fails to abandon her animal ways. 'A barn, even a stable, would have suited me fine, but on my own, alone. I was still grunting in my sleep […] Honoré had no reason to keep me.'[21] I am reading Pig's vulnerable position and maltreatment in light of Agamben's 'Homo Sacer', a being who has broken the law, a figure who causes disorder in a community, but

---

19   The passage is from the introduction of Marie Darrieussecq's *Zoo* (Paris: P.O. L., 2006) p. 7. It is translated from French by Anat Pick, in *Creaturely Poetics: Animality and Vulnerability in Literature and Film* (New York: Columbia University Press, 2011), p. 81.

20   Following the shooting at *Charlie Hebdo*'s offices in 2015, Darrieussecq offered to write for the magazine, supporting its freedom of speech and its feminist character, as well as its critical stance against racism. Rosemary Neill, 'Charlie Hebdo terror attack galvanised Marie Darrieussecq into action,' *The Australian*, April 13, 2016.
      <http://www.theaustralian.com.au/arts/review/charlie-hebdo-terror-attack-galvanised-marie-darrieussecq-into-action/news-story/94b78b5391a4cb6e84bd6c01aee6d2dd > [accessed 5 March, 2017]

21   Darrieussecq, p. 36.

still cannot be sacrificed, or eaten in Pig's case.[22] The protagonist is constantly persecuted by the authorities in both her human and animal form, either for her murderous actions or her meat. Drawing upon the Roman concept of sacredness according to which a criminal's life was 'sacred' and therefore not suitable for sacrifice, Agamben defines his 'sacred' man as part of the political and social life, especially because a *homo sacer* has been denied the life and the rights of a citizen (*bios*), he only has 'bare life' (*zoe*), that is a life not governed by human laws.[23] Asibong's argument that Pig's 'bare life is filtered through sex' demonstrates the political implications of Agamben's *homo sacer* concerning women, but also powerless beings who, like Pig, are other than human.[24] Pig's political existence is essentially outside human law because her criminal and oversexualised instincts unsettle the rules of the Parisian community. As Asibong observes, 'the Law's perversely sexual agenda can be subverted in the cultivation of such intense sacredness that new political identities may occur.'[25] The Law's most nightmarish side manifests itself in a dystopic version of Paris in the novel: work conditions are terrible, especially for women, totalitarian governments are hostile to foreigners, mentally ill people, stray animals and women, all personal information is monitored and projected in mass media, while books are banned alongside critical thinking, freedom and animal rights. As Sanja Bahun has argued on Asibong's interpretation of Agamben, 'Darrieussecq significantly expands the circle of victims' to men, animals and other liminal beings.[26] Therefore, the *mulier sacra* can meaningfully function within a newly negotiated political existence but only if the space is inclusive of similarly fluctuating beings. Reading the woman-sow as a rejected member of the society may highlight an aspect of hybrid subjectivity: abjection. The term, proposed by Kristeva, encompasses the parts of our self that we cannot contain and

---

22   Giorgio Agamben, *Homo Sacer: Sovereign Power and Bare Life*, trans.by Daniel Heller-Roazen, rev. edn. (Stanford: Stanford University Press, 1998).
23   Agamben, *Homo Sacer*, p. 60.
24   Asibong, p. 171.
25   Asibong, p. 177.
26   Sanja Bahun-Radunovic, 'The Ethics of Animal-Human Existence: Marie Darrieussecq's *Truismes*', in *Myth and Violence in the Contemporary Female Text: New Cassandras*, eds. Sanja Bahun Radunovic and Julie Rajan (Surrey: Ashgate, 2011), pp. 55-74 (p. 66).

often disgust us, such as dirty skin.[27] In Darrieussecq's novel, the pink pig skin which emerges as part of the transforming body is unbearable for both the subject and her surroundings. 'It was only after I put on a little too much weight that I began to disgust myself,' confesses Pig, while later on she discovers that bathing is efficient against casting off the animal urges and skin changes.[28] Following Kristeva's argument on the various aspects of abjection, I suggest that Darrieussecq's protagonist inhabits the 'fragile state' Kristeva describes, where human meets the animal. By eventually embracing her abject animalistic and oversexualised side, the pig protagonist can qualify for an unbearable existence which bridges the distance with the monstrous Other.[29]

The political significance of such an existence goes beyond an individual's exchange with a farm animal though. It is useful to read, for instance, Pig's seemingly allegorical experiences, and her inability to recognise them, in contrast with the reader. After the election of a conservative, and eventually oppressive, political party in France, animals almost disappear and most books are censored. While hiding in a mental institution, Pig comes across Knut Hamsun's novel *Hunger* which has been banned.[30] She then reads a passage, on animal killing:

> Then the knife plunges in. The farmhand gives it two little shoves to push it through the thick skin, after which the long blade seems to melt through the neck fat as it sinks in up to the hilt. At first the boar doesn't understand a thing, he remains stretched out for a few seconds, thinking about it. Aha! Then he realizes he is being killed and utters strangled cries until he can scream no more.[31]

She is unable to understand the allegorical meaning of it, though; even more difficult it is for her to understand the implications of the passage for animals, like herself. 'I wondered what a *boar* was; my back began to feel all clammy. I decided to laugh it off, because

---

27   Julia Kristeva, *Powers of Horror: An Essay on Abjection*, trans. by Leon S. Roudiez (New York: Columbia University Press, 1982), p.53.

28   Darrieussecq, p. 16.

29   Kristeva, *Powers of Horror,* p. 54.

30   Paradoxically, the quote is not from Hamsun's novel *Hunger*, but from his 1908 novel *Benoni*, as Anat Pick has observed (2011). Darrieussecq quotes the same passage as an epigraph in *Pig Tales*, not mentioning where it is taken from. It seems to me that the use of the title *Hunger* is ironic, yet an evidence of postmodern literature.

31   Darrieussecq, p. 87.

otherwise I was going to throw up',[32] she confesses. Here Darrieussecq approaches allegorical tales with irony, since allegories do not appeal to animals or hybrid beings, even if they are understood by readers. In fact, a postmodern standpoint on allegories and metaphors is attempted; they are stripped bare to the basics of meaning. The animal change of a woman into a pig can easily be read as a metaphor for the unfair treatment of women and animals in postmodern societies. Yet Darrieussecq goes beyond the prescriptive discourse on stories of metamorphosis as an opportunity to discuss pressing social issues that are primarily anthropocentric. Literally read, as Bahun suggests, the pig woman is a political being in her hybridity.[33] Anat Pick adopts a similar approach to the text, considering the exiled being as probing 'the permanent interval of species, the trembling space between the human and the animal.'[34] We can then read Darrieussecq's story as a narrative about animal-human subjectivity which does not favor the human only, in this case women, but rather advocates an equal exchange between species. Read from this perspective, the trope of metamorphosing from human to animal can be said to function as a platform on which contemporary society can redefine their treatment of animals. This non-metaphorical character of the metamorphosis is useful for the conceptualization of transformations into which animality is taken seriously and is even capable of providing animals' own narratives.

Loss, Melancholia, and the Act of Writing through the Event of Metamorphosis

Is it possible for a pig to write? This is how Darrieussecq's porcine protagonist responds to this unusual question:

> I know how much this story might upset people, how much distress and confusion it could cause. But I must write this book without further delay. Simply holding a pen gives me terrible cramps. I hope that any publisher patient enough to decipher these piggle-squiggles will

---

32 Darrieussecq, p. 87.
33 Bahun-Radunovic, p. 69.
34 Anat Pick, *Creaturely Poetics: Animality and Vulnerability in Literature and Film* (New York: Columbia University Press, 2011), pp.79-99 (p. 89).

graciously take into consideration the enormous effort I'm making to write as legibly as possible.[35]

The reader is warned from the very first line that the story about to follow is an unsettling one and that 'any publisher who agrees to take on this manuscript will be heading for trouble.'[36] Darrieussecq weaves the fictional, first-person narrative of a hybrid woman-sow creature that looks back to the events that brought her to her present condition, in the form of a quick paced autobiographical narrative written in a state of urgency. The illusion of the urgent and hasty aspect of this confession is also marked by the lack of chapters in the novel; it is written, instead, in the form of a long, continuous text. The events that led the anonymous narrator into a peculiar state of writing 'piggle-squiggles' unfold when, desperate for a job, she starts working as a salesgirl and sex worker in a cosmetics chain in Paris. It is during this period that she starts gradually transmogrifying into a sow. As if speaking retrospectively, the narrator now understands that her change should not have been entirely surprising to her, had she been more observing to the symptoms. Her human flesh, also tested by the chain director before she hires her, was 'marvelously elastic' while her thighs 'had grown pink and firm, curvaceous, yet muscular.'[37] In a relentless narration of squealing, blood-letting dreaming, abortions, flower eating, and futile attempts to disguise the emerging pig with makeup products, the narrator discovers another self in her. She has become more animal-like, not only on the outside, and she is therefore not welcome in contemporary Parisian society, especially during a far-right regime. After the regime changes into a seemingly freer one, the pig-woman meets another being who can occasionally transform himself like her: Yvan, the former owner of the cosmetics company *Moonlight Madness*, who is a werewolf.[38] While they live happy together, continuously moving between their human and animal form, the Society for the Protection of Animals which operates in Paris, kills the wolf. The encounter with Yvan is a

---

35 Darrieussecq, p. 1.
36 Darrieussecq, p. 1.
37 Darrieussecq, pp. 2-3.
38 Yvan's company is literally called 'Loup-Y-Es-Tu', a word-play in French which hints at the shifting nature of the company owner but also at the supposedly transformational-yet unsuccessful, as we have already seen in the text- power of the cosmetics industry. Also, Marie de France's *Lais* 'Bisclavret' resonates in Darrieussecq's werewolf narrative.

transforming experience for Pig, and it is his tragic death that led her to write her adventures in the first place.

The image of a porcine writer may seem less lyrical, when compared with other animals, such as Colette's artistic nightingale in her story 'The Tendrils of the Vine' (1908),[39] in which the author's creativity is compared to the nightingale's voice, yet also emerging from a near death experience. Julia Kristeva emphasises that in Colette the parallelisms to the bird are 'the *very realization* of the change under way' and in fact are 'better than metaphors, they are metamorphoses,'[40] an interpretation which is in line with my reading of the relationship between a metamorphosed, non-metaphorical subject and creativity in *Pig Tales*. The urge to write in order to establish and communicate a sense of identity is encouraged by the protagonist's inexplicable metamorphoses. Importantly, this identity is conceived as hybrid. The pig-natured woman is able to record both the experiences of an underprivileged woman and of a farm animal but she gains a deeper understanding of herself only when she accepts both states of being; she also learns to empathise with realistically marginalised identities within French society, such as African people, Muslim women and migrant workers.

Kristeva has also examined another component of the writing and suffering schema: sadness. In *Black Sun* Kristeva posits melancholia in a linguistic context, as language, but also as an 'affect', a mood manifested due to traumas.[41] The body, besides the psyche, responds to sadness over a loss and literary creation emerges out of this response, when the affect of sadness is exteriorized. 'Literary creation is that adventure of the body and signs that bears witness to the affect – to sadness as imprint of separation and beginning of the symbol's sway', argues Kristeva.[42] Signs and symbols emerge and are identified by the subject with the loss of a loved object, which is often maternal.[43] These symbols are not stable though; the 'sway' means that the melancholic writer doubts the very symbols she is using, thus there is a constant questioning of the power of language to convey the experienced sadness. Kristeva warns that the final literary creation, 'the testimony' as she calls it, will not be identical

---

39   *The Collected Stories of Colette*, ed. by Robert Phelps, trans. by Herma Briffault (New York: Farrar, Straus, Giroux, 1983).
40   Kristeva, *Colette*, p.96.
41   Kristeva, *Black Sun,* p. 22.
42   Kristeva, *Black Sun*, p. 22.
43   Kristeva acknowledges Hanna Segal's position on the necessary condition of loss for creative activity to emerge. Kristeva, *Black Sun*, p. 23.

to the sadness which created it, although the reader will be able to identify the 'semiotic and the symbolic imprints' of the affect.[44] Pig's language is not fully understood by humans or it is often misunderstood, especially when she displays human-like emotions. 'Look at that! She's crying!', exclaims one guest at a New Year's celebration during which Pig is ready to be slaughtered but she feels grateful for the food instead of being afraid.[45] In another scene, Pig is again offered food as she is homeless but she is unable to utter human speech: 'I tried to thank him but couldn't speak clearly!'[46] In her discussion of literary metamorphoses, Kristeva argues that 'narrative does not suit the metamorphic body, which is located beyond the footlights, is without barrier and without prohibition' thus differentiating between the ordered human communication and a story which comes from a metamorphosed being, a human-animal, so to speak.[47] Indeed, Pig's vulgar and explicit language and the emergence of a non-human language are humorously blamed on her transformation. Kristeva highlights that an 'ek-static jubilation lies behind a poetics of the fragment in which [...] a woman, a man, or a cat assume the same importance.'[48] As Kristeva revisions depression and melancholia as affecting the work of art, both in a productive and a destructive way, it is worth noting that in theorising creativity and sadness she emphasises the continuous dialogue between melancholia and writing but also the circumstance after sadness has dissolved. The text functions now as a 'fetish' allowing for associations of writing with pleasure, a feeling that our Pig protagonist enjoys throughout her narration of her adventures.[49] In what follows, I will read Pig as an emerging writer who finds her voice after a loss, and a melancholic subject. Kristeva's views on maternal loss and the emergence of literature further assist me in identifying the construction of hybrid subjectivity in *Pig Tales*. Writing, for the pig protagonist, may function as a mourning mechanism for the loss of a maternal loved object.

It is the wolf's death, another hybrid creature, which triggers the work of art, Pig's confessional narrative. 'Yvan loved me equally well as a woman and a sow. He said it was fantastic to have two modes of being

---

44   Kristeva, *Black Sun*, p. 22.
45   Darrieussecq, p. 93.
46   Darrieussecq, p. 61.
47   Kristeva, *Colette*, p. 97.
48   Kristeva, *Colette*, p. 199.
49   Kristeva, *Black Sun*, p. 9.

for the price of one.'[50] Pig becomes aware of her peculiar complexity only after her encounter with Yvan who teaches her how to coexist in-between two spaces. Thus, the act of writing in the novel both assists the mourning work and engenders a more comprehensive transformation; the fantasy of physical metamorphosis is thematically and semantically linked to this inner transformation of the protagonist. For this reason, while there are metamorphic changes, or changes of identities, there is narration; when these changes cease, the narration ends. After the death of perhaps the only being that could understand the peculiar complexity of her nature, Pig suffers both in her human form and her porcine entity. The pig recalls: 'they put me in a van and then in a cage at the zoo. I screamed for several days. I wasn't eating. I lay down to await death.'[51] But it is noteworthy to underline that Darrieussecq's bereaved pig is also unusually gifted; since she is terrified that she might submit to an animalistic nature only, which to her would mean losing herself, she tries to stand up in two legs and remember what is like to be human. The protagonist's need for self-expression initially stems from the loss of her werewolf-mate, and at the end of the story it becomes imperative for her acceptance of her own hybrid nature. Her mourning, expressed through writing, allows her human self to come to light and thus being able to write. The admission 'that's why I write; it's because I remain myself through my sorrow over Yvan'[52] testifies that the ability to find yourself in writing (perhaps literally here) is a privilege of humans. In Darrieussecq's story the act of writing, which is writing about one's self in this case, is vital to preserve the human memories as it allows the protagonist to reflect upon both her past and present state in relation to her identity. What additionally preserves her human memory and body is her memory of Yvan; his memory is what makes her to stand in two legs while she is still a pig which is about to be slaughtered by her own mother.

Kristeva identifies an additional element that she sees as the ultimate trigger of melancholia, apart from the love-object of identification which Freud had described in his 1917 essay,[53] and she terms this Ur-

---

50   Darrieussecq, p. 108.
51   Darrieussecq, p. 124.
52   Darrieussecq, p. 128.
53   Kristeva's work is largely indebted to Freud. In 'Mourning and Melancholia'
     Freud defines both melancholia and mourning in terms of the loss of a love-
     object. The object loss is 'transformed into an ego-loss' thus causing
     ambivalence to the melancholic person due to the identification with a lost
     object. Freud adds a third 'precondition' to melancholia, besides the loss of

Element the 'Thing'. The 'Thing' is not an object though, but a pre-oedipal 'non-object of desire and loss that escapes signification'[54] which cannot be named, represented or signified; hence it becomes difficult for the melancholic writer to successfully mourn for it.[55] Kristeva suggests that this non-object resides somewhere in the realm of the maternal. The loss of the maternal symbolic object leads the melancholic into wishing to compensate for its loss, hence to respond to this loss in a specific, imaginative way which Kristeva considers representative of a new way of mourning. Writing has often been linked with the writer's inner psychological state; and in Kristeva's thinking and her analysis of literary examples it has a healing function against melancholia. The statement that 'there is no imagination that is not, overtly or secretly, melancholy'[56] can be therefore used to identify the role of writing as a powerful mourning ritual in Darrieussecq.[57] The 'Thing' in Kristevan theory represents the mother; hence male subjects (and writers) are able to bear this symbolic loss. The case is different for females, argues Kristeva, since for a woman to discard the maternal object would mean to reject a part of her own self, to commit a symbolic matricide, thus sacrificing the part of herself which had been identified with the mother.[58] Matricide is necessary for the female artist to emerge out of the melancholy position; if she fails to discard the maternal Thing, the writer finds herself in the state of 'asymbolia' which leads to 'the death of the self,' thus condemned into silence and invisibility.[59] Kristeva's views on matricide and female discourse have been often criticised because of the lack of alternatives

---

the object and ambivalence that is 'the regression of libido into the ego.' Sigmund Freud (1917), 'Mourning and Melancholia', in *The Standard Edition of the Complete Psychological Works of Sigmund Freud, Volume XIV (1914-1916): On the History of the Psycho-Analytic Movement, Papers on Metapsychology and Other Works*, trans. by James Strachey (London: The Hogarth Press and the Institute of Psycho-analysis, 1957), pp. 237-258.

54 Tsu-Chung Su, 'Writing the Melancholic: The Dynamics of Melancholia in Julia Kristeva's *Black Sun*,' *Concentric: Literary and Cultural Studies* 31.1 (2005) 163-191, p. 167.

55 Kristeva, *Black Sun*, p. 13.

56 Kristeva, *Black Sun*, p. 6.

57 My purpose is not to use psychoanalytic criticism to interpret Marie Darrieussecq in terms of the process of writing her own private phantasies. I am reading *Pig Tales* in the light of the above theorists, and I am focusing on how the protagonist becomes a writer.

58 Kristeva, *Black Sun*, p. 28.

59 Kristeva, *Black Sun*, p. 28.

she offers for feminine discourse.[60] It seems to me though that Kristeva's views on maternal loss and the emergence of literature bring to light significant aspects of female subjectivity in *Pig Tales*.

After reconnecting with her mother, Pig remains in her mother's farm, always oscillating between woman and animal. In a tense episode, Pig regains a human shape for a few moments and manages to murder her mother to avoid being slaughtered for her meat. This episode can be interpreted as a symbolic act of separating herself from the maternal space thus confirming what Kristeva has argued in *Black Sun*: 'matricide is our vital necessity, the sine-qua-non condition of our individuation.'[61] Darrieussecq's protagonist's killing of her mother is significant as the mother has ironically paired with Pig's former exploiter, and they are getting ready to sell her meat in the black market until Pig manages to shot them.[62] Female writers, argues Kristeva, need to commit a symbolic matricide, thus sacrificing that part of themselves which had been identified with the mother. Matricide is necessary for the female artist to emerge out of the melancholic position. Almost playfully engaging with Kristeva's text, and literalising its metaphors, Darrieussecq suggests that by reckoning with the loss of the mother – by a libidinal 'matricide'– one can find the viable alternative to the self-destructive tendencies and is enabled to act creatively, to utilise signs and symbols, and to write creatively. Significantly, though, it all has to happen at the place of the maternal itself. Indeed, the anonymous female protagonist of *Pig Tales* takes refuge in a farm which belongs to her mother. It is the same farm where she returns to steal a notebook so she can record her story. 'Maternal instinct is a wonderful thing, a gut feeling, as they say' thinks the pig upon meeting her mother.[63] By hinting the existence of the maternal as unconscious or inherited, Darrieussecq has her protagonist to reject the mother and supposedly to accept the loss of the Thing theorised by Kristeva. Pig is not then a melancholic because she denies the death of the mother; she has accepted it.

After the murder, it is strongly implied that the pig can acquire a human form in full moon: 'when I crane my neck towards the Moon, it's to show, once again, a human face.'[64] At the end of Darrieussecq's

---

60   See for example Tsu-Chung Su (2005).
61   Kristeva, *Black Sun*, p. 36.
62   Darrieussecq, pp. 133-134.
63   Darrieussecq, p. 120.
64   Darrieussecq, p. 135.

story, another transformation is visible in the sow-writer, this time caused by the very act of writing. The woman-sow at the end of the novel is different from the one at the beginning because she is fully aware of her hybrid self and body. The ideal writer, Darrieussecq seems to suggest in her novel, has to incorporate a hybrid identity in order to be able to write her story. Borrowing Virginia Woolf's words, she has even acquired, 'a room' or, rather, a farm of her own, a space wherein her hybrid subjectivity can be written and meaningfully accommodated.[65] Darrieussecq stretches the concept of a room into a place of female creativity, thus this space becomes literally and metaphorically larger than the four walls of the room, and perhaps more abstract than before: it could be a site. A place of writing can be anywhere, even in an abandoned farm or a forest, and can still develop and probe a compensating fantasy of female writing. The sow-woman combination is now balanced and less confusing for the subject: 'I'm a sow most of the time. […] I watch television. I phoned the director's mother. […] I'm not unhappy with my lot.'[66] She has now acquired a place of her own, a space wherein her hybrid subjectivity can be inscribed.

Hybrid Subjectivity: Animal or Feminine Writing?

The 'second' becoming of the Pig creature is that of a writer, who has to remain human in order to write. The Pig's final words that 'I write whenever my animal spirits subside a little. The mood comes over me when the Moon rises, and I reread my notebook in its cold light' testify that a type of écriture informs Pig's subjectivity as much as her animal instincts.[67] Becoming a writer is equally illuminating for the protagonist and the reader: for Darrieussecq, it is the female writer who has to embrace a wider realm of disenfranchised beings, those who are also mistreated and disadvantaged by patriarchy. Hélène Cixous has addressed the connections between female discourse and the female body in her essay 'The Laugh of the Medusa' (1971) and

---

65 By reaffirming the need for women to have their own space for writing, Marie Darrieussecq's most recent translation of Woolf's celebrated essay 'A Room of One's Own' in 2016 serves as a literary reminder that this space is not yet large enough. Marie Darrieussecq, trans. *Un lieu à soi*, Virginia Woolf (Paris: Éditions Denoël, 2016).

66 Darrieussecq, pp. 134-135.

67 Darrieussecq, p.135.

her insistence on reclaiming a female literary territory undoubtedly accommodates some of Darrieussecq's concerns on the formation of female subjectivity. Cixous vividly suggests that every

> woman must write her self: must write about women and bring women to writing, from which they have been driven away as violently as from their bodies-for the same reasons, by the same law, with the same fatal goal.[68]

As women have been repeatedly silenced, embarrassed and excluded from speech by men, the only means for them to enter the phallocratic symbolic order is to write about their unconscious desires, wishes and thoughts, and the events surrounding the female body, which has also been demonized in male discourse. Pig's uncensored account of the female self and body stresses the need of a female mode of expression which is autonomous from society's laws. Her transformations are foremost somatic and coexist or alternate with intimate, and often taboo, female processes. Pig's transformation is, for instance, in tune with her menstruation: 'I found it hard to get used to my body's new rhythm. I got my period about every four months, following a short phase of *sexual excitement*, not to mince my words.'[69] Social change is embedded in women's writing according to Cixous, as female writing can be 'the very possibility of change, the space that can serve as a springboard for subversive thought, the precursory movement of a transformation of social and cultural structures.'[70] In a similar vein, for Darrieussecq, literature should aspire 'to put words where there are no words, to put words where we don't know what to say and even less what to write.'[71] Incorporating social change in women's writing may assist us in reading the pig woman's writing as such; hybrid-female subjectivities can mobilise changes in the established male discourse which is treated ironically in Darrieusseq's text. Pig's autobiography is not strictly a product of female writing, but also an écriture de cochon, a product of

---

68   Hélène Cixous, (1971), 'The Laugh of the Medusa' trans. by Keith and Paula Cohen, in *New French Criticisms* ed. by Elaine Marks and Isabelle de Courtivron (New York: Schocken Books, 1981), pp. 245-267.
69   Darrieussecq, p. 35.
70   Cixous, p. 879.
71   'The Grand Budapest Hotel, Marie Darrieussecq; Alan Ayckbourn,' interview, BBC Radio 4, March 3rd, 2014. < http://www.bbc.co.uk/programmes/b03wpdy8> [accessed 5 March 2017]

marginalization and exile from civilization. The trope of metamorphosis in this equation assists the reader to understand how hybridity is created, and why subjectivity matters in such texts which advocate female writing.

Facing the Horror of neither Animal nor Human

Through my reading of Kristeva and Cixous, I have been emphasizing the dialectic between loss and writing in Darrieussecq's abject animal being. I have argued that it is precisely Pig's exiled status which can accommodate the shocking narrative of a woman being transformed into a pig. I have considered creative writing in *Pig Tales* as a transformational experience for female subjectivity, and at the same time a subjectivity which is itself metamorphic and linked with significant body changes in a woman's body. With *Pig Tales* Darrieussecq communicates the fantasy of a modern female writer with no education or financial security, of dubious morals and overt sexuality who manages to write her autobiography based on her own transformations, primarily bodily ones, which at the very end of the narrative are also mental. It is a 'hybrid' writer who can open up to the possibility of human-animal coexistence, both somatic and mental, and can engage with the pressing issue of how we define the human as opposed or coexisting with the animal in a stimulating narrative.

Returning to both Agamben and Kristeva's understandings of excluded subjectivities, and by acknowledging the fear and suffering these narratives stress when it comes to our entrance to the animal territory, it seems to me that these twentieth-century responses still carry the trauma of our coexistence with other species. Yet, considering the horror of Pig confronted with the prospect of remaining an animal at the beginning of the novel, and her acceptance of her hybrid nature through the act of writing her autobiography, it is imperative that such carnivalesque narratives of horror and humour are told, for they allow us to negotiate our own identity and its abject components, one of them being the animal. Darrieussecq's novel seems to suggest that the hierarchical way in which society treats women is comparable to the way in which it treats animals, but also that both feminine humans and animals have some special gifts that can be realised not so much by opposing but going beyond the prescriptive discourse on women and animals. The vacillation of the Pig is not seen as an escape from

a dystopic society but as a creative, even terrifying, opportunity for a
dialogue between different subjectivities.

Bibliography

Agamben, Giorgio, *Homo Sacer: Sovereign Power and Bare Life*, trans. by
    Daniel Heller-Roazen (Stanford: Stanford University Press, 1998)
Agamben, Giorgio, *The Open: Man and Animal*, trans. by Kevin Attell
    (Stanford: Stanford University Press, 2004)
Asibong, Andrew, 'Mulier Sacra: Marie Chauvet, Marie Darrieussecq and
    the Sexual Metamorphoses of 'Bare Life'', *French Cultural Studies*, 14.2,
    (2003), 169-177
Bahun-Radunovic, Sanja, 'The Ethics of Animal-Human Existence: Marie
    Darrieussecq's *Truismes'* in *Myth and Violence in the Contemporary
    Female Text: New Cassandras*, eds. Sanja Bahun-Radunovic and Julie
    Rajan (Surrey: Ashgate, 2011), pp. 55-74
BBC Radio 4, 'The Grand Budapest Hotel, Marie Darrieussecq; Alan
    Ayckbourn,' interview, March 3, 2014 < http://www.bbc.co.uk/programmes/
    b03wpdy8> [accessed 5 March 2017]
Burgess, Glyn S., and Busby, Keith (trans.) Marie de France, *The Lais of Marie
    de France* (London: Penguin Classics, 1999)
Cixous, Hélène, (1971), 'The Laugh of the Medusa' trans. by Keith and
    Paula Cohen, *New French Criticisms* ed. by Elaine Marks and Isabelle de
    Courtivron (New York: Schocken Books, 1981), pp. 245-267
Cox, Fiona, trans. Marie Darrieussecq interview, (2012) in *Practitioners' Voices
    in Classical Reception Studies, Special Issue 2013: Contemporary Women
    Writers* < http://www.open.ac.uk/arts/research/pvcrs/2013/darrieussecq>
    [accessed 5 March 2017]
Damlé, Amaleena, *The Becoming of the Body: Contemporary Women's Writing
    in French* (Edinburgh: Edinburgh University Press, 2014)
Darrieussecq, Marie, *Truismes* (Paris: P.O.L, 1996)
Darrieussecq, Marie, *Pig Tales: A Novel of Lust and Transformation*, trans.by
    Linda Coverdale (London: Faber and Faber, 1997)
Darrieussecq, Marie, *Zoo* (Paris: P.O.L., 2006)
Darrieussecq, Marie, trans. Virginia Woolf, *Un lieu à soi* (Paris: Éditions
    Denoël, 2016)
Freud, Sigmund, (1917), 'Mourning and Melancholia', *The Standard Edition
    of the Complete Psychological Works of Sigmund Freud, Volume XIV
    (1914-1916): On the History of the Psycho-Analytic Movement, Papers
    on Metapsychology and Other Works*, trans. by James Strachey, ii-viii,
    (London: The Hogarth Press and the Institute of Psycho-analysis, 1957),
    pp. 237-258
Jordan, Shirley, *Contemporary French Women's Writing: Women's Visions,
    Women's Voices, Women's Lives, Vol. 37*, (New York: Peter Lang, 2004)

Kafka, Franz, 'The Metamorphosis', trans. by Susan Bernofsky, intr. by David Cronenberg, 1st ed. (New York: W. W. Norton, 2014)

Kristeva, Julia, *Powers of Horror: An Essay on Abjection*, trans. by Leon S. Roudiez (New York: Columbia University Press, 1982)

Kristeva, Julia, *Black Sun: Depression and Melancholia*, trans. by Leon S. Roudiez (New York: Columbia University Press, 1989)

Kristeva, Julia, *Colette*, trans. by Jane Marie Todd (New York: Columbia University Press, 2004)

Muirden, Sallie, 'Magical Allegory in Marie Darrieussecq's Pig Tales (1996): Piggy debauchery in postcolonial France', *COLLOQUY text theory critique*, 16 (2008), 231-244

Neill, Rosemary, 'Charlie Hebdo terror attack galvanised Marie Darrieussecq into action', *The Australian*, April 13, 2016 <http://www.theaustralian.com.au/arts/review/charlie-hebdo-terror-attack-galvanised-marie-darrieussecq-into-action/news-story/94b78b5391a4cb6e84bd6c01aee6d2dd> [accessed 5 March, 2017]

Ovid, *Metamorphoses*, trans. by A. D. Melville, rev. edn. (Oxford: Oxford University Press, 2008)

Pick, Anat, *Creaturely Poetics: Animality and Vulnerability in Literature and Film* (New York: Columbia University Press, 2011), pp.79-99

Phelps, Robert (ed), *The Collected Stories of Colette*, trans. by Herma Briffault (New York: Farrar, Straus, Giroux, 1983)

Tsu-Chung, Su, 'Writing the Melancholic: The Dynamics of Melancholia in Julia Kristeva's *Black Sun*,' *Concentric: Literary and Cultural Studies* 31.1 (2005) 163-191

Warner, Marina, *Fantastic Metamorphoses, Other Worlds: Ways of Telling the Self* (Oxford: Oxford University Press, 2002)

Marie Cazaban-Mazerolles

ST. MAWR, LASSIE AND A 'NAIVE HEDGEHOG': THEORETICAL AND PRACTICAL ISSUE ABOUT REPRESENTING THE ANIMAL *QUA* ANIMAL IN LITERARY NARRATIVES

Western literature has a long history of both symbolization and anthropomorphization of non-human animals. From Aesope's fables to Orwell's revolutionary pigs; from medieval bestiaries to Lewis Carroll's bunny; or from Baudelaire's albatross to Leopardi's *passero solitario* – literary animals more often than not 'appea[r] as significant figures [...] strictly in terms of metaphor'.[1] Yet over the last century, a growing number of writers in line with the global reassessment of animal nature and our human relationships with them have opposed such a use of animal figures, and ostensibly strived to reach a non-allegorical, non-symbolical representation of the animal as an animal. Like human sciences, literature is 'now struggling to catch up with a radical evaluation of the status of nonhuman animals that has taken place in society at large'.[2]

Through the analysis of British writer D.H. Lawrence's short story 'St. Mawr' (1925), which features a young lady rejecting the society she lives in in the aftermath of her encounter with a powerful stallion and was considered by Margot Norris as one of the earliest attempts to 'restor[e] the animal *qua* animal to literature by liberating it from its tropological enslavement to the human',[3] this paper first contends that such a way to frame the issue is problematic since it relies upon a homogeneous view of 'the animal' that reproduces dualism and ultimately runs the risk to lead back to the anthropocentrism it was supposed to escape. I will then turn to the French contemporary writer

---

1   Susan McHugh, *Animal Stories: Narrating Across Species Lines* (Minneapolis: University of Minnesota Press, 2011), p.7.

2   Cary Wolfe, *Zoontologies* (Minneapolis: University of Minnesota Press, 2003), p.xi.

3   Margot Norris, *Beasts of the modern imagination. Darwin, Nietzsche, Kafka, Ernst, and Lawrence* (Baltimore & London: John Hopkins University Press, 1985), p.18.

Éric Chevillard's *Du hérisson* (2002), in which a writer who is about to pen his autobiography confronts a hedgehog suddenly appeared on his desk, as an example of how the use of specific ethological and ecological knowledge can provide a way to solve this first difficulty. By drawing on contemporary ethological debates, I will nonetheless try to unveil new anthropocentric biases in such representation. Finally, the rewriting of collie-star Lassie's tale by the contemporary American writer T.C. Boyle in 'Heart of a champion', a short story published in 1974 in which the relationship between the young Timmy and his faithful dog is seriously amended, will provide the opportunity to question the relationships between animal individualization and anthropomorphism as the ultimate pitfall any literary representation of the animal has to deal with.

It must be immediately added that such an argumentative agenda does not aim at assessing literary works on a pass-or-fail basis, and shall not be considered as a normative discourse distinguishing between good and bad representations of the literary animal. The reader then will be asked to accept the fact that the literary texts under consideration will, to some extent, undergo a test they are not intended to pass. One should recognize the full legitimacy of literary works to feature animals for purposes other than naturalistic *mimesis,* and keep in mind that literary discourse is not to be evaluated according to what contemporary scientific knowledge and cultural trends claim is true or false, right or wrong.

Therefore, the paper's aim is rather to enlighten the innovative paths explored by 20th and 21st centuries narratives in order to create a less anthropomorphic or anthropocentric literary representation of the animal, without assuming that such an ambition stands for the authors' whole poetic project. In addition, the framework within which literary scholars engage with the so-called animal question will be put under scrutiny, so that not only the writers' representations but also the general theoretical and lexical biases will be critically examined.

Lawrence's 'St. Mawr': an all too Platonic Animal

In her book *Beasts of the Modern Imagination*, Margot Norris indicates Lawrence as an early representative of what she terms a biocentric tradition that originated from Nietzsche and Darwin. Lawrence's short novella 'St. Mawr', she states, especially attests to the British author's inclination towards a non-anthropocentric, non-anthropomorphic, and non-symbolical representation of the animal:

He treats the encoding of the animal as symbol, metaphor, or allegory, as an impoverishment and a denigration, and […] he restores the animal *qua* animal to literature by liberating it from its tropological enslavement to the human.[4]

I agree to a large degree with such a claim. Indeed, the horse St. Mawr, eponymous character of the story, is mostly shown as a figure who resists symbolization despite the numerous attempts by the human characters who surround him. Whether it is Lou (the horse's owner), her husband, her mother, or the upper-class London people: each and every character keeps trying to give the stallion a metaphorical significance. In their discourses, it sometimes appears as God, sometimes as Evil, sometimes as an allegory for rebellion from those who are treated like slaves, or even as a symbol of sexuality. Nonetheless, the very proliferation of such attempts and the contradictions they give rise to, none of them being endorsed by the narrative voice, ultimately prove them wrong. St. Mawr resists all these conflicting guesses which Lawrence appears to mock, so that the endeavor of symbolization itself is bound to fail. As shrewdly noted by Norris, the human proclivity to give non-human animals symbolical significance is disclosed in the story as a new kind of exploitative power and violence. After St. Mawr overthrows Lou's husband Rico during a ride, the symbolical interpretation of its behavior leads to a death sentence:

> But St. Mawr? Was it the natural wild thing in him which caused these disasters? Or was it the slave, asserting himself for vengeance?
> If the latter, let him be shot. It would be a great satisfaction to see him dead.[5]

Nonetheless, I believe a few restrictions should be added to Norris' statement. First, it is clear in the novella that 'the liberation [of the animal] from its tropological enslavement' is achieved at the cost of its outright disappearance from the text. Towards the end of the story indeed, St. Mawr who has been brought to the United States by her mistress starts showing signs of interest for mares, compelling the human characters to acknowledge it as a bodily creature filled with instincts which are

---

4   Norris, p.17-18.
5   David Herbert Lawrence, *St Mawr and others stories* (Cambridge: Cambridge University Press, 1983), p.82.

directed towards its own species. The horse turns into a horse and ceases to be an idea, thus inhibiting any symbol-making impulse: 'And St. Mawr followed at the heels of the boss's long-legged black Texan mare, almost slavishly. What, in heaven's name, was one to make of it all?'.[6] Hence this is the moment when the horse, not being able to serve as a support for Lou's or anyone's projections anymore (nothing can be 'made' of it), disappears from the story: 'They left St. Mawr and Lewis'.[7]

Such disappearance at the time when Lou has just left both her husband and Britain in order to meet with the wild American West only makes clearer the overall diegetic function of the horse: St. Mawr is left behind because it has completed its mission which was to awake Lou to an alternative way of life opposed to the one lived by the refined men and women of the modern society who repress the animal within themselves. Such a role then necessarily stems from a process of essentialization which presents St. Mawr as a mere embodiment of the abstract idea of 'the animal'. So, even though Lawrence's story does question the long tradition of projecting anthropomorphic features and all-too-human fantasies onto non-human animals, I contend that it doesn't disrupt as much as it replicates a tropological use of the animal figure since St. Mawr ultimately appears to stand as a synecdoche for animality. The animal remains a symbol, but one which takes place in an autotelic allegory: here, the animal character refers to an undefined animality, without any attention being paid to the singularity of its own way of being an animal.

In this respect, Lawrence's discourse is to be held responsible for the same sort of idealization that Lou shows when she tells her mother: 'I don't want to be an animal like a horse or a cat or a lioness, though they all fascinate me, the way they get their life *straight*'.[8] We see here that there is no consideration at all for the differences that exist between the life of a horse and the life of a lion, Lou acknowledging only one uniform 'straight' animal life. According to Philip Armstrong, here lies a typical modernist bias which prevents the de-symbolization process to be fully achieved:

> Where industrial modernity reduces animals to a collection of raw materials or a sequence of processes, modernist aesthetic sublimates

6   Ibid., p.132.
7   Ibid.
8   Ibid., p. 61.

them into essence. [...] Clearly, then, a particular animal can only ever provide a temporary signifier of the redemptive power of vital 'animality'.[9]

This is what I propose to call – by drawing on a note about Ted Hughes' poetics made by Coetzee's fictional character Elizabeth Costello in *The Lives of Animals*[10] – the Platonic bias in the sense that the animal character is left to embody an abstract Platonic Idea of animality. And so does Norris, somehow, when she talks about 'representing the animal *qua* animal', referring to an essentialist idea of what is an animal, despite the now-famous warning by Derrida about using the name 'the animal', in its singular form:

> [A] notion as general as 'the Animal', as if all nonhuman living things could be groups without the common sense of this 'commonplace', the Animal, whatever the abyssal differences and structural limits that separate, in the very essence of their being, all 'animals', [is] a name that we would therefore be advised, to begin with, to keep within quotation marks. Confined within this catch-all concept, within this vast encampment of the animal, in this general singular, within the strict enclosure of this definite article [...], are *all the living things* that man does not recognize as his fellow, his neighbors, or his brothers. [...]. There is no animal in the general singular, separated from man by a single indivisible limit. We have to envisage the existence of 'living creatures' whose plurality cannot be assembled within the single feature of an animality that is simply opposed to humanity.[11]

In Derrida's view, the monolithic concept of 'the Animal', or the very Platonic idea of 'animality', eludes the heterogeneity and diversity of animal lives and then reactivates the dualism in which the thesis of human exceptionalism originates. Consequently, it can be said that every mention (or representation) of 'the animal' is loaded with

---

9   Philip Armstrong, *What Animals Mean in the Fiction of Modernity* (London & New York: Routledge, 2008), p.149.
10 Commenting on Hughes' poem 'The jaguar', Costello states that 'despite the vividness and the earthiness of the poetry there remains something Platonic about it' (Coetzee, p. 53). Incidentally, it should be stressed that Costello begins her analysis by putting Hughes 'in a line of poets who celebrate the primitive' (Ibid., p.52) among which she puts Lawrence himself.
11 Jacques Derrida, 'The Animal That Therefore I Am (More to Follow)', trans. by David Willis, *Critical Inquiry*, 28.2 (2002), 369-418 (pp. 402; 415).

anthropocentrism since it acknowledges only the unicity of the human being, whereas *homo sapiens* doesn't have the monopoly of being unique and each and every animal singularity demands recognition.

Chevillard's Hedgehog: towards Eco-etho Specification

This is the same path that the French writer Éric Chevillard seems to follow in *Du hérisson*, in which he escapes such a Platonic bias by paying accurate attention to the ecological and ethological specificities of the animal he features.

The first paragraph of the book foreshadows the effort of singularization and witnesses a particularly interesting use of the phrase 'the animal' criticized above.

> Cela m'a tout l'air en effet d'un hérisson naïf et globuleux, l'animal, là, sur mon bureau. Je ne crois pas me tromper. J'ignore comment il est arrivé ici, ou qui l'y a mis et pourquoi. Que dois-je en faire ? [...] Je connais mal cet animal, je l'avoue, le hérisson naïf et globuleux / ne m'est pas familier.[12]

The abstract significance of the formula 'l'animal' used in the very first sentence is here annihilated by its use as an apposition to the preceding nominal group 'un hérisson naïf et globuleux', that stipulates which specific kind of animal the narrator is dealing with and undercuts any Platonic bias thanks to the use of the indefinite determiner, further reinforced through the use of the deictic reference 'ici'.[13] Hence at the end of the paragraph, 'l' animal' is replaced with the more specific 'cet animal', thus attesting to the fact that the text evades both abstraction and essentialization. Though the narrator confesses he does not have a clue about hedgehogs in the last sentence, the text is about to show how he overcomes such ignorance and how he tries to gain access to the singular and specific life of this animal being he has just met.

---

12   Éric Chevillard, *Du hérisson* (Paris : Minuit, 2002), p. 9. 'It does seem to be a naive and fuzzy hedgehog, the animal, here, on my desk. I don't think I'm wrong. I don't know how it got here, nor who brought it nor why. What should I do with it? [...] I don't know this animal very well, I have to confess, the naive and fuzzy hedgehog / is unfamiliar to me.' (All translations are mine).

13   I will comment later on the two adjectives 'naïve and fuzzy' that systematically qualify the noun 'hedgehog' throughout the text.

As announced by the book's title derived from an old-fashioned structure the author borrows from scientific discourse, natural sciences function as a major source for Chevillard's narrative which regularly offers its reader scientific insights about hedgehogs, their way of life, their diet, their natural habitat, and so on. We learn for instance that their average life expectancy is between seven and eight years, that 'la saison reproductive du hérisson naïf et globuleux dure d'avril à août',[14] and that 'il nage, il n'aime pas l'eau mais il nage bien, et vite'.[15] Drawing on specific ethological, ecological and anatomical knowledge, Chevillard's representation of the animal thus avoids the abstract discourse about 'the animal' and subsequently the inherent anthropocentrism it conveys.[16]

However, I will contend that this text, while efficiently opposing the essentialization of the animal, expresses another tropological bias that I would call the Aristotelian one starting from the French philosopher Baptiste Morizot's recent work *Les Diplomates,* where he makes the following statement:

> Notre compréhension infrastructurelle de l'animal est restée aristotélicienne. On croit que l'essence (les traits spécifiques) suffit pour étudier le comportement d'un animal: que l'individu n'excède pas les traits spécifiques.[17]

---

14 Chevillard, p.47. 'the breeding season of the naive and fuzzy hedgehog lasts from April to August'.

15 Ibid., p.92 : 'he swims, he doesn't like water but he's a good and fast swimmer'.

16 Stating that paying attention to ecological and ethological specificities of an animal is a way to avoid its abstract representation does not presume the necessity of true etho-ecological features. The important point here is to provide the animal character with specific singularities, with no regard for their scientific relevance. In this respect, Jim Crace's representation of *pseudogryllidus pelagicus* in *Being dead* - a fictional species of sea-crickets that is granted in the novel with very specific, if necessarily made-up, ethological and ecological features - can be said to avoid abstraction and idealization as much as Chevillard's conscientiously realistic hedgehog.

17 Baptiste Morizot, *Les Diplomates* (Marseille: Wildproject, 2016), p.133. 'Our infrastructural understanding of animals has remained an Aristotelian one. We are still convinced that in order to study the behavior of an animal, there is no need to look any further than its essence (the characteristics of the species): that the individual doesn't exceed its specific features'. My translation.

Chevillard's hedgehog embodies hedgehogness: it is no more than one exemplar of the species it belongs to, expressing only features which are specific to its species, without being singularized as an individual. The text does not grant it any particular or individual (as opposed to specific) characterization, so that it still appears as a genre of synecdoche: this hedgehog stands for all hedgehogs. Even the 'naïve' and 'fuzzy' adjectives used all along the book and likely to individualize the hedgehog are said to be a taxonomic qualifier – the narrator ensuring the reader he is only naming a characteristic of the whole species here: 'Je suppose que le hérisson naïf et globuleux doit ce qualificatif taxinomique de *naïf* à son regard principalement'.[18]

Consequently, Chevillard's hedgehog does not stand as a proper individual, nor as its narrative equivalent, that is, as a proper character. And in such reluctance to grant animals individuality, one can find again an anthropocentric residue that denies the very possibility of a genuine animal subject – subjectivity being considered as another privilege of the one and only human being. Therefore, as stated by Erica Fudge in 'Reading animals', fully eschewing anthropocentrism implies the representation of a truly individualized animal:

> [T]he truly meaningful animal is often a very individualized being. That is, whatever the intellectual context of discussion –religious, legal, scientific and so on- it is often the singular animal –*that* sheep, rather than general sheep- that has the greatest power to upset human status.[19]

At this point, it is fundamental to add two different remarks about Chevillard's 'failure' to represent animal individuality and subjectivity in his text. Firstly, even though this may not be a deliberate goal of the author, one can notice that the hedgehog, (prevented from acquiring a genuine subjectivity by the narrator's limited literary representation) precisely turns out to prevent the human narrator and protagonist of the tale from writing his autobiography, that is from building his own representation as an individual subject by the means of a writing performance. In *Du hérisson*, the animal figure as well as the human narrator fail to achieve subjectivity through narration, as if the animal, thwarting the human writer's commitment to give a particularized

---

18   Chevillard, p.19. 'I assume the naïve and fuzzy hedgehog essentially owes the taxonomic qualifier naïve to its gaze'.
19   Erica Fudge, 'Reading Animals', *Worldviews*, 4 (2000), 101-13 (p.110).

and singular representation of himself, was paying him back for the a-biographical nature of human discourses about animals.

Secondly, such state of things is all the more ironic considering that, whereas Chevillard has chosen to draw on the genre of scientific treatises as the main intertext for his tale thus subsequently depriving himself of the particularizing virtues of the literary narrative form throughout the making of characters, contemporary ethological sciences are now promoting literary narrative forms as a highly relevant hermeneutic pattern. As stated by the French philosopher Dominique Lestel:

> Une des grandes découvertes de l'éthologie de ces trente dernières années est d'avoir démontré qu'il est difficile de décrire et comprendre certains animaux sans faire l'hypothèse qu'ils sont des individus, c'est-à-dire que rendre compte de ce qu'ils sont passe par une biographie qui excède largement la description comportementale. [...] Ces animaux ont une *histoire*.[20]

The American bio-ecologist Marc Bekoff and the bioethicist Jessica Pierce called for a 'narrative ethology' in their book entitled *Wild justice: the Moral Lives of Animals*, by praising stories as a means to 'stimulate thought, activate the imagination of scientists, lead to new questions, represent anomalies, and challenge conventions of thought';[21] whereas the French philosopher and wolf behavior's expert Baptiste Morizot also demands a new narrative epistemology likely to oppose the old-fashioned Aristotelian one that, he states, "[s'est] trop longtemps concentré[e] sur ce qu'il y a de *commun* à toute une espèce [...] supposant trop peu de variabilité des animaux'.[22] By contrast, he argues that it is necessary to pay attention to the importance of storytelling in the production of individual difference: 'Il faut ici narrer le processus

---

20 Dominique Lestel, *L'Animal singulier* (Paris: Seuil, 2004), p. 36-37. 'One of the most important discovery made in ethology for the last thirty years was the demonstration that we can't describe and understand some animals without assuming that they are individuals. In other words, to give a full account of what they are implies a biography that goes far beyond the mere behavioral description. [...] These animals have a *story*.' My translation.

21 Marc Bekoff and Jessica Pierce, *Wild justice: the Moral Lives of animals* (Chicago & London: University of Chicago Press, 2009), p. 37.

22 Morizot, p. 133. 'ethology has focused on what is *common* to a whole species for too long, [...] assuming too little variability among animal individuals'.

des événements, l'histoire des variations sur ce qu'on croyait être la norme'.[23]

In other words, narration, being the place where variations and singularities can be told, forces animal individuality into our speeches – something which Chevillard's old-fashioned ethological model (with Buffon being his primary source) could not achieve. The strength and potential of the literary narrative form – when it comes to unsettle the anthropocentric bias it contributed to champion for such a long time – are ultimately brought out into open. But then an old question rises again: to what extent can the individualization of the animal figure in the form of a literary character eschew the anthropomorphization typical of centuries of Western literature's representation? And can anthropomorphism *per se* not be the real issue?

Boyle's Lassie: the Issues of Anthropomorphization and Autonomy in the Representation of Animal Subjectivity

The case of Lassie, the famous fictional Collie character firstly imagined by the British writer Eric Knight who then became a Hollywood star through the eponymous American TV show, is emblematic of the endeavor consisting in providing an animal figure with genuine individuality and subjectivity. Throughout its adventures, the dog grew as a highly individual character, granted with a peculiar psychological profile and an exceptional intelligence going far beyond the features usually attached to its breed. According to psychology professor Stanley Coren in his best-selling and awards-wining *The Intelligence of Dogs: A Guide to the Thoughts, Emotions, and Inner Lives of Our Canine Companions*, Lassie's impact on the way we might refer to non-human animals was great:

> We believed that this dog (thus, by extrapolation, all dogs) could think, plan, sympathize, feel pain, have emotions of sorrow and joy, remember complex facts, and even plan acts of retribution. Hadn't we actually seen Lassie do it?[24]

---

23   Ibid., p. 149. 'One needs to tell the stories of dynamic events, of the variations departing from what we once thought was the norm'.

24   Stanley Coren, *The Intelligence of Dogs: A Guide to the Thoughts, Emotions, and Inner Lives of Our Canine Companions* (New York: Simon and Schuster, 2006), p. 10.

Yet T.C. Boyle's rewriting of the story in 'Heart of a Champion' shows how such an individualization through narration and the construction of the character led to both anthropomorphization and anthropocentrism again, since all too human features, and more importantly all-too-human desires were projected onto the dog character.

Lassie meets with sponsor - 1954.   Lassie and Tommy - circa 1955.

Boyle's short story moves along a series of scenes described as if they were looked at by an external viewer putting both narrator and reader together through the use of the pronoun 'we'. Such configuration makes it clear that his text was not designed as a rewriting of Knight's novel, but as a remake of Robert Maxwell's TV show, whose excessive anthropomorphism is targeted. Indeed, as the Collie starts accomplishing its usual tricks and feats, preventing young Timmy boy to get smashed by a falling trunk as early as the second paragraph, Boyle soon gives the dog character an overly human behavior, to the extent that Lassie 's skills as 'Man's best friend' just appear farfetched and ludicrous. For instance, not only is Lassie able to rescue its master from the roaring waters he has fallen into, but it also gives him mouth-to-mouth resuscitation: 'The collie sniffs at Timmy's inert little form, nudges his side until she manages to roll him over. Then clears his tongue and begins mouth-to-mouth'.[25] Later on, when Timmy has been knocked unconscious by another falling trunk, the reader witnesses Lassie rushing home, where interspecies communication appears not to raise any difficulty at all:

---

25   T.C. Boyle, *Descent of Man* (Boston & Toronto: Atlantic Monthly Press Book, 1974), p. 39.

'What is it girl? What's the matter? Where's Timmy?'
'Yarf! Yarfata-yarf-yarf!'
'Oh my! [...] Timmy's trapped under a pine tree out by the old Indian
burial ground –'
'Arpit-arp.'
'– a mile and a half past the north pasture.'[26]

Here, Boyle 's irony towards the televisual representation of the dog
character slammed for its unrealistic anthropomorphism is obvious.
However, I shall argue that Boyle 's short story displays a more complex
way to set up such an issue – and makes a subtler statement about it.

Because of its similarities with *The Call of the Wild*, it might be
argued that 'Heart of a Champion' is a rewriting not only of Robert
Maxwell's show but also of Jack London 's story. That Boyle was a
keen reader of London, whose name frequently pops up in his works and
interviews, is an established fact. Surely, one can assert that Lassie going
through the same pattern as Buck – from domestication to responding
to the 'call of wild' here embodied by a coyote instead of a wolf –
is no coincidence. Yet *The Call of the Wild*, published in 1903, was
one of the targeted books during what is known as 'the Nature Fakers
controversy', inaugurated by the naturalist and writer John Burroughs
who rebuked the early-century American writers' tendency to picture
the animals in a sentimental, unscientific and anthropomorphized way
in their novels. In 1908, London puts an end to his previous silence
and writes a harsh response to both Burroughs and President Roosevelt
who had joined the debate in the meanwhile. Referring to both *The Call
of the Wild* and *White Fang*, the novelist defends his representation of
dogs' reasoning and feeling as typical of dogs' nature, and as confirmed
by the Darwinian evolutionist theory.[27] He then dismisses the very
accusation of anthropomorphism as a 'homocentric' one, which denies
any 'kinship with the other animals' by holding even simple reasoning
and feelings as exclusively human features.[28]

---

26   Boyle , p. 43.
27   Darwin's notebook M contains many observations on dogs, granted with
     feelings such as shame, pride and, to some extent, free-will by the British
     naturalist.
28   Jack London, 'The other animals', *Revolution and other Essays* (New York:
     The Macmillan Company, 1910) p. 259. As judiciously noticed by Hub
     Zwart (*Understanding Nature. Case Studies in Comparative Epistemology*,
     p.122) London 'elaborates a line of argument here that will be taken up later

Such a background invites to look back suspiciously at 'Heart of a Champion': is really T.C. Boyle, an admirer of London and a fervent supporter of Darwinian Theory, taking on himself to duplicate the old guard's argument about animal stories? It is unlikely. Actually, Lassie is granted with inner thoughts till the end of the text in Boyle 's story, that is, even when it is not Maxwell's character anymore, but rightfully Boyle 's. Moreover, Boyle's own additions to the archetypal Lassie scenarios do not shy away from anthropomorphism, as displayed in the seduction game between the Collie and the coyote that echoes *Romeo and Juliet*'s balcony scene:

> [S]he rises and slips to the window, silent as a shadow. And looks down the long elegant snout to the barnyard below where the coyote slinks from shade to shade […]. [The coyote] leers up at the window and begin a soft, crooning, sad-faced song.[29]

Consequently, it appears that if Boyle 's short story does sneer at excessive anthropomorphism, his main target is not the picturing of an animal as a conscious and sensitive subject – which was Burrough's accusation – as much as shaping a complacent, obedient, and ultimately heteronomous animal subjectivity.

Indeed, dog-star Lassie's subjectivity is marked by heteronomy on at least three different levels. Firstly because the dog character, called a 'sentimental icon'[30] by Henry Jenkins, has been granted with a personality that would fit the values American society was seeking to celebrate during the fifties and sixties, that is, obedience and heroism. As stated by Kelly Wolf, Lassie's adventures featured 'a nostalgic longing for the traditional values of loyalty and working-class pride lost on the onset of modernity and industrialization'; the dog protagonist figuration being 'instrumental in constructing appropriate standards of obedient behavior and American heroism'.[31] As a cultural icon, Lassie's character is shaped according to the needs and desires of the American

---

by Freud himself in his famous essay 'Eine Schwierigkeit der Psychoanalyse' (1917). Scientific breakthroughs such as the theory of evolution contain a 'narcissistic offense'.

29   Boyle, p.40.
30   Henry Jenkins, *The Wow Climax: Tracing the Emotional Impact of Popular Culture* (New York: New York University Press, 2007), p. 222.
31   Kelly Wolf, 'Promoting Lassie. The animal star and constructions of 'ideal' American heroism', *Cinematic canines: Dogs and their work in the Fiction*

society of its time. Secondly, the very values it is meant to exemplify make it a complacent animal, whose every action and thought are directed towards its human companions. Boyle's story makes this two-fold human alienation very clear in his text, displaying how the representation of nature in the show ought to gratify both human spectators and characters. Here is how the first paragraph of the novella reads:

> We scan the cornfields and the wheat fields winking gold and goldbrown and yellowbrown in the midday sun [...]. There'd have to be a breeze – and we're not disappointed [...]. The boy stops there to gaze out over the nodding wheat [...]. Then he brings three fingers to his lips in a neat triangle and whistles long and low [...]. And then we see it –way out there at the far corner of the field – the ripple, the dashing furrow, the blur of the streaking dog.[32]

The representation of nature is as complacent to the human viewer (giving them the breeze they expect) as Lassie is obedient to Timmy (answering his whistle). Even the wheat is 'nodding' here, since Boyle's problem is not the anthropomorphic gesture *per se* but rather the obedience model from which it does not depart. Lastly, as a TV character, Lassie was played by a real dog – Pal – who was performing under Maxwell's staff, that is, acting just what it was asked to act and whose image was strictly controlled. Lassie, Boyle notices in his story, has 'her lashes mascaraed and curled'.[33]

On the three levels, the representation of Lassie 's subjectivity is thus shaped by human others (American society, Timmy and his family, Pal's trainers and marketing team) following human desires and rules. By contrast, Boyle 's concern is precisely to emancipate his character and to grant it with an *autonomous,* not humanly-carved, subjectivity.

Therefore, his dog character progressively becomes less and less compliant with the young Timmy, to the extent that Lassie leaves its young master die and flees with its new coyote friend which turns out to be instrumental in Lassie's metamorphosis. Concomitantly, the text displays a more and more untamed and uncensored poetics, indifferent to the viewer's sensibility. For instance, when Lassie meets with the coyote

---

     film, ed. by Adrienne L. McLean (New Brunswick: Rutgers University Press, 2014), p.106.

32   Boyle, pp. 37-38.

33   Ibid., p. 39.

in the woods, it chooses not to fight it – so that Timmy and his family's poultry are safe – but rather engages in sexual intercourse with it:

> [S]he licks at his whiskers, noses at his rear, the bald black scrotum. Timmy is horror-struck. Then, the music sweeping off in birdtrills of flute and harpstring, the coyote slips round behind, throat thrown back, black lips tight with anticipation.[34]

Not only does the scene allows the animal figure to escape the cheap sentimentalism of the TV show, but it also displays the dog's achievement of its own pleasure, so that its actions as a character are not bound to human beings anymore, but are now self-directed. Lassie ceases to be an instrumental subject, while its inner life breaks the hold of human beings. At some point, the text says, '[w]e watch the collie's expression alters in midbound – the look of offended AKC morality giving away, dissolving'.[35] The change of gaze expresses the conversion of the dog's subjectivity. Boyle's story thus frees Lassie not only from its domestic way of life but also from the status of sentimental anthropocentric icon that Hollywood assigned to her.[36] Providing the animal with inner-life and the capacity to think and feel is not enough to make it a genuine subject; indeed, it needs also to be an independent 'I' able to build its own identity against heteronomous human discourses.

The ultimate irony, however, is that the full completion of such representation of the animal as an autonomous subject demands its disappearance, so that it will escape even the (human) narrator's and the reader's gaze. Therefore, at the end of the story, Lassie leaves Timmy behind in the valley where he is about to drown:

> The two animals start at that terrible rumbling, and still working their gummy jaws, the dash up the far side of the hill. We watch the white-tipped tail retreating side by side with the hacked and tick-blistered grey one – wagging like ragged banners as they disappear into the trees at the top of the rise.[37]

---

34   Ibid., p. 41.
35   Ibid., p. 44. The American Kennel Club (AKC) is a registry of purebred dog pedigrees in the United States.
36   In 1960, the character of Lassie received a star on the Hollywood Walk of Fame and in 2005, the show business journal *Variety* named Lassie one of the '100 Icons of the Century'.
37   Ibid., p. 45.

Lassie is not a 'good girl' anymore, but an animal among other animals, beyond the reach of all-too-human discourses.

Conclusion

Since the beginning of the twentieth century, Western fictional narratives have endeavored to cleanse the literary representation of the animal from its symbolical, metaphorical and, broadly speaking, anthropocentric *habitus*, and have moved towards picturing it as an animal *per se*, freed from the instrumentalization it has been subjected to for such a long time. Nonetheless, this paper has tried to disclose three biases causing the subsistence of both a tropological use of the animal and anthropocentrism, including in our own critical and theoretical discourses.

Firstly, the *Platonic* bias, which consists in representing the animal so that it merely embodies an abstract, ideal, and homogenous idea of animality, which thus becomes a symbol within an autotelic allegory, as illustrated by Lawrence 's novella. Secondly, the *Aristotelian* bias, when attention is paid to ecological and ethological characteristics of the animal without recognising any kind of individuality beyond the features of its own kind, so that one given animal appears to stand for the whole species it belongs to, a tendency exemplified by Chevillard 's book. Finally, the *Pygmalion* bias, which consists in not being able to picture animal subjectivity in and for itself, but only patterned after an all-too-human perspective, that is shaped by human desires, will, rules, and standards so that the animal character is granted with a deficient iconic subjectivity that acknowledges its individuality, but not its autonomy, an issue addressed in Boyle 's short story.

All three biases in a certain way, and to some extent, duplicate rather than defeat the anthropocentrism they challenged in the first place. Therefore, one may legitimately ask whether the issue of the representation of the animal '*qua* animal' is not ultimately an aporetic one.

In this respect, it is worth noticing that in the three texts examined, the animal emerges as a resistant figure defying even its own literary representation. St Mawr remains ontologically alien by escaping Lou's understanding, resisting attempts to be trained, and finally disappearing from the text long before the story ends. So does Lassie by leaving the scene with its coyote mate and giving up on its human masters; whereas Chevillard 's hedgehog remains an obscure character, mostly an obstacle that prevents the narrator from writing, eats his papers, and in a word, is an embarrassment: 'Que dois-je en faire?'[38] the narrator asks repeatedly in the opening lines of the book. Each time, a certain kind of adversity characterizes the relationship between the animal and the human being who is trying to figure out how to proceed with it. And each time, it seems that the genuinely successful way out of anthropocentrism ultimately consists in the detachment and remoteness of the animal figure, finally eluding the very possibility of human speech and representation.

Following Donna Haraway's analysis of photographic safaris as a continuation of the hunter's violence in which the camera replaces the gun and forces the animal to 'hold forever the gaze of the meeting',[39] or even Steve Connor's extension of the same theme to arts – when he argues that every representation that aims at making animals conspicuous is a violation since 'most natural creatures exist in a kind of discretion, semi-concealment'[40] –, one might argue that allowing the animal to hide within the literary text and to escape our gaze as readers is finally the ultimate and paradoxical liberation the author can provide. This perhaps explains the recurrence, noticed by the French literary scholar Anne Simon, of the representation of the animal as a hiding character, 'un être de fuite' whose main phenomenological manifestation within twentieth and twenty-first century literature proves to be elusiveness. Yet the fact remains, as she puts it, that 'plus l'animal fuit, plus l'écrivain le cherche'.[41]

Bibliography

Armstrong, Philip. 2008. *What Animals Mean in the Fiction of Modernity* (London & New York: Routledge).
Bekoff, Marc and Pierce, Jessica. 2009. *Wild justice: the Moral Lives of animals*

---

38   Chevillard , p. 9. 'What should I do with it?'
39   Donna Haraway, 'Teddy bear patriarchy: Taxidermy in the Garden of Eden. New York City, 1908-1936', *Social Text*, 11 (1984-5), 20-64, (p.25).
40   Steve Connor, 'Such Stuff As Dreams Are Made On', expanded version of an essay published as 'The Right Stuff', *Modern Painters*, 2009, 58-63 (online, np).
41   Anne Simon, 'Chercher l'indice, écrire l'esquive: l'animal comme être de fuite, de Maurice Genevoix à Jean Rolin', *La question animale*, ed. by Jean-Paul Engélibert et al. (Rennes: Presses Universitaires de Rennes, 2011), p.170. 'The more the animal flees from him, the more the writer tries to track it down'. My translation.

(Chicago & London: University of Chicago Press).

Boyle, T.C. 1974. *Descent of Man* (Boston & Toronto: Atlantic Monthly Press Book).

Chevillard, Éric. 2002 (2012). *Du hérisson* (Paris, Minuit).

Coetzee, John Maxwell. 2009. *The lives of animals* (Princeton: Princeton University Press).

Connor, Steve. 2009. 'Such Stuff As Dreams Are Made On', expanded version of an essay published as 'The Right Stuff', *Modern Painters*: 58-63. Online.

Coren, Stanley.2006. *The intelligence of dogs: A Guide to the Thoughts, Emotions, and Inner Lives of Our Canine Companions* (New York: Simon and Schuster).

Crace, Jim. 1999. *Being Dead* (New York: Farrar, Strauss and Giroux).

Derrida, Jacques. 2002. 'The Animal That Therefore I Am (More to Follow)', trans. by David Willis, *Critical Inquiry* 28.2: 369-418.

Fudge, Erica. 2000. 'Reading Animals', *Worldviews* 4:101-13.

Haraway, Donna. 1984-1985. 'Teddy bear patriarchy: Taxidermy in the Garden of Eden. New York City, 1908-1936', *Social Text* 11: 20-64.

Jenkins, Henry. 2007. *The Wow Climax: Tracing the Emotional Impact of Popular Culture* (New York: New York University Press).

Lawrence, David Herbert. 1983. *St Mawr and others stories* (Cambridge: Cambridge University Press).

Lestel, Dominique. 2004. *L'Animal singulier* (Paris: Seuil).

London, Jack. c1931. *The Call of the Wild: and other stories* (New York: Macmillan).

London Jack. 1910. 'The other animals', *Revolution and other Essays* (New York: The Macmillan Company).

McHugh, Susan. 2011. *Animal Stories: Narrating Across Species Lines* (Minneapolis: University of Minnesota Press).

Morizot, Baptiste. 2016. *Les Diplomates* (Marseille: Wildproject).

Norris, Margot. 1985. *Beasts of the modern imagination. Darwin, Nietzsche, Kafka, Ernst, and Lawrence* (Baltimore & London: John Hopkins University Press).

Simon, Anne. 2011. 'Chercher l'indice, écrire l'esquive: l'animal comme être de fuite, de Maurice Genevoix à Jean Rolin', *La question animale*, ed. by Jean-Paul Engélibert et al. (Rennes : Presses Universitaires de Rennes): 167-181.

Wolf, Kelly. 2014. 'Promoting Lassie. The animal star and constructions of 'ideal' American heroism', *Cinematic canines: Dogs and their work in the Fiction film*, ed. by Adrienne L. McLean (New Brunswick: Rutgers University Press): 104-120.

Wolfe, Cary (ed.). 2003. *Zoontologies* (Minneapolis: University of Minnesota Press).

Zwart, Hub. 2008. *Understanding Nature. Case Studies in Comparative Epistemology* (Berlin: Springer Verlag).

VALENTINA SAVIETTO

BOTHO STRAUß: MYTHOLOGY AND SYMBOLOGY WITHIN THE CONCEPTION OF A 'HUMAN ANIMALITY'

Introduction

This essay will focus on the issue of mythology and symbology in the theatre and prose works of the German dramatist and writer Botho Strauß (1944). The analysis will focus on two main texts, i.e. the play *Schlußchor* ('Final Chorus', 1991) and the early novel *Der junge Mann* ('The Young Man', 1984, engl. 1995). Both works present a vast panorama of animal representation, as well as a revolutionary anthropological conception; therefore, it is possible to outline that in Botho Strauß the borders between humanity and animality are and become absolutely permeable and ambiguous. The essay will analyse on the one hand the 'animal' use of language, i.e. metaphors as well as mythological images, which are strongly connected to the linguistic domain of animal studies; on the other hand, this text will examine the author's poetics within the research field of animal studies and will underline the narrative processes through which Strauß's characters come to a progressive transformation from their human nature into the animal one. The following article aims at demonstrating that animal representation within contemporary German literature, – here especially considering the example of Botho Strauß, – opens up the possibility to debate fundamental questions about the conception of human beings as well as their anthropological status.

Postdramatic Theatre and Animal Studies

On 1st February 1991, the director of the 'Münchner Kammerspiele' Dieter Dorn put on the first representation of Botho Strauß's political drama *Schlußchor*, facing through this play not only the question of postdramatic theatre, but also the very topical subject of the fall of the Berlin Wall as well as the issue of the German reunification, i.e. the *Wiedervereinigung*. At that time the author, essayist and dramatist

Botho Strauß had already been acknowledged as an important German writer, especially after having been awarded the 'Georg-Büchner-Preis' in 1989, although the question of his national value as a poet was still strongly influenced by the political division of the 'Two Germanys' – as is generally known, Botho Strauß was born in Naumburg, in the state of Saxony-Anhalt, which belonged to the German Democratic Republic, but studied in western universities as Cologne und Munich. For this reason, Dorn's decision to stage Strauß's new play *Schlußchor* is to be interpreted as a political and cultural gesture, since the dramatist, in spite of his undisputed talent, was regarded as a much-discussed intellectual, often self-effacing and prone to scandalous intellectual positions, as shown by his early controversial theatre or literary works *Trilogie des Wiedersehens* ('Trilogy of Return' – literally 'Trilogy of seeing again', 1977), *Die Widmung* ('The Dedication', 1977) and *Paare, Passanten* ('Couples, Passersby', 1981, engl. 1996).

In this context, it is clear that Dorn intended to offer to his reunified country an early representation of the *Wiedervereinigung* topic on stage, but obviously under a dramatic perspective, or rather under the postdramatic one. In this essay, the idea of postdramatic theatre represents a key concept, because its main features, i.e. fragmentation, plurality and media scepticism, are often highly present also in those literary works in which the animal motif strongly emerges. For this reason, postdramatic theatre could also be understood as a way and a *medium* of representation, illustration and explanation of important matters within the research field of animal studies.

The way in which the postmodern vision and the 'animal' perspective of the Animal Studies melt in *Schlußchor* can be explained by considering the formal conception of this *pièce* as well as the characters playing in its three acts. According to the principles of postdramatic theatre, the acts are seemingly disjointed from each other and function as three independent plays within a unique play. However, these are in any case connected to each other through thematic nuclei which rely on fundamental narrative motifs, especially the animal and human ones.

Individuals vs Mythic Figures in Schlußchor

In order to deal with the animal theme in Botho Strauß's poetics, as illustrated in the play *Schlußchor*, it seems particularly interesting to consider first of all not its opening, but its central act *Lorenz vor dem*

Spiegel (Aus der Welt des Versehens) ('Lorenz in front of the mirror (From the world of the oversight)'), which focuses on the destiny of the title character, as well as on the above-mentioned object, the mirror. In this central part of the play, the main character, an aesthete and architect, utters a strange statement which concerns his image reflected in the mirror. The man looks at himself through this object and deals with it as if it were a living being too: '*Er wendet sich zum Spiegel. Der Spiegel schweigt und mustert einen unbegreiflichen Menschen*' ('*He turns to the mirror*. The mirror remains silent and scrutinizes an incomprehensible man').[1]

In the context of Animal Studies, both presence and reference to this 'mirror matter' are strongly connoted, since they immediately recall a central theoretical reference within this research domain, i.e. Jacques Derrida's *L'animal que donc je suis* ('The Animal That Therefore I Am', 2006, engl. 2008).[2] Similarly to the French philosopher's theory, the mirror image of *Schlußchor* is an important narrative device too, which brings to the fore the central issue of the social, political, psychological and emotional boundaries among species, in other words between human and animal beings.

Schlußchor's mirror precisely evokes the neutral, upright and therefore higher position of that famous cat, which faces the philosopher Derrida, naked and fully ashamed because of this condition. For instance, it is possible to identify a kind of nakedness also in the character of Lorenz: even though he is dressed – while attending a very elegant party in a German villa –, he is metaphorically 'undressed', or, more precisely, lacking an essential human feature, language. This detail becomes clear by observing his behaviour in front of the mirror, which can be qualified as clumsy and eccentric. Lorenz appears many times in the dressing room or in the toilet, where the mirror is, with the purpose of trying and improving his love declaration to the beautiful woman Delia, a client of his, so that he rehearses his speech precisely in front of the mirror. However, it should be highlighted that, at this point of the *pièce*, Lorenz has already lost his job at Delia's house. He should have renovated her apartment, but, after an unpleasant accident, he lost the job. Even in this case, Lorenz's attitude is to trace back to the sense of sight, so that the presence of the mirror can be interpreted as the thematic development

---

1   Botho Strauß, *Schlußchor* (München: dtv, 2007), p. 49. [My translation]
2   See Jacques Derrida and David Wills, 'The Animal That Therefore I Am (More to follow)', *Critical Inquiry*, 28, No. 2 (2002), 369-418.

of the above-mentioned accident, from whose description the second act of *Schlußchor* starts inter alia and which involves, not by chance, precisely the sight theme.

Botho Strauß's text plays with the semantic field of the sight, the gaze, the glance, but also the fortuitous vision, so that the act of seeing is immediately presented as a guilty attitude. In the first scene of this second act, Lorenz enters Delia's apartment without giving his name, as he is convinced that the woman is still on her way home. On the contrary, she is already at home, actually taking a bath, and Lorenz catches a sight of her, naked while bathing. Therefore, this accident is due to an act of mistake through sight, in German *Versehen*, and immediately arouses in Lorenz's soul a strong sense of guilt. As a consequence, human look is presented by Botho Strauß not as impartial, but as strongly connoted, and differs for this reason not only from the neutral position of the mirror in the second act, but above all from the superior cat's glance in Derrida's essay.

Consequently, the sight motif in Botho Strauß's work is strictly linked to the anthropological definition of human behaviour. That becomes especially clear by analysing Delia's reaction to the gesture of the architect. She perceives his act not as unintentional, but as responsible, as this intentionality is precisely what marks out humanity. Derrida himself discusses this problem in his lecture *L'Animal que donc je suis*: according to the French philosopher, human beings feel modesty, shame and guilt. For this reason, humans are basically different from animals, or rather, they are not just and simply 'animals', but they put an air of superiority or they assume ambiguous positions, as humans can imitate animals too, as the French verb underlines (*je suis* means both 'I am' and 'I follow'). Following and going after are phenomena which evoke an important lexical domain within Animal Studies, i.e. the field of hunt, seduction, predation and eroticism, all features which are largely present in *Schlußchor* and which involve both sexuality and death.[3]

In the act *Lorenz vor dem Spiegel* for instance, the fact that the artist has spied the naked woman while bathing triggers an unconscious mechanism of predation, whose development follows a religious and ritual pattern, i.e. the relationship between David and Bathsheba, but also a mythological one, as demonstrated by the strong mythic foundation of

---

3   Cf. Gianfranco Dalmasso, 'Introduzione all'edizione italiana: Il limite della vita', in Jaques Derrida, *L'animale che dunque sono*, ed. by Marie-Louise Mallet and Gianfranco Dalmasso (Milano: Jaca Book, 2006), pp. 7-28 (p. 21).

all Strauß's works.[4] In this case, the mythological subject refers to the ancient myth of Diana and Actaeon, which tells the unfortunate story of the young hunter Actaeon, as he sees the goddess of the hunt, naked while enjoying a bath in a spring helped by different nymphs. For his involuntary act of seeing he is thus punished and transformed into a deer, unable to speak and to express himself through human language. His destiny ends with his violent death, as he is torn by his own hounds, which are now unable to recognize their master. In a similar way, Lorenz is no more able to utter comprehensible sounds or words in front of his beloved; as a consequence thereof, he is demoted to the 'animal' rank and is punished – just as Actaeon – with death, though he suicides.

In this context, remarkably, the author charges this act with a great symbolic value: if on the one hand, Lorenz just surprises a woman having a bath, on the other hand dressing and being undressed become the evident sign of the confrontation between a man, seeking after himself and the 'truth', and a woman symbolising exactly this 'truth', which corresponds to the idea of 'naked truth' of Derrida's notes related to *L'Animal que donc je suis*.[5] Lorenz's progressive inability to establish a relationship with Delia is depicted in *Schlußchor* by a gradual 'animal metamorphosis' on the part of the architect, through which the text also indicates that Lorenz is more and more deviating from his research of the truth. Besides, this purpose is at the same time a will to self-determination:

> LORENZ Ich habe sie... ich habe sie vor mir!... Erwischt! Schnell! Wie sagt man denn, wie heißt es treffend: 'was mir vollkommen undurchsichtig... was mir völlig schleierhaft – gänzlich nebulös›? Ein Meer von Varianten! Ich stehe wie gelähmt vor dem Reichtum meines Deutschs! Muttersprache! Deine Hand!... Undurchsichtig? Ach, Spiegelatur! Rede, wie dir der Schnabel gewachsen ist!... Ist mir ein Schnabel gewachsen? Undurchsichtig... [...] Un-durch-sichtig-... Kreuzun-durchsichtig-... Kreuzschnabel...[6]

---

4   Cf. Sigrid Berka, *Mythos-Theorie und Allegorik bei Botho Strauß* (Wien: Passagen-Verl., 1991). See Eva C. Huller, *Griechisches Theater in Deutschland: Mythos und Tragödie bei Heiner Müller und Botho Strauß* (Köln u.a.: Böhlau, 2007), pp. 349-364.

5   Cf. Derrida and Wills, 369.

6   *Schlußchor*, p. 47. The German term *Spiegelatur*, lexically closed to *Spiegel* ('mirror'), was originally used by the German mystic Jacob Böhme in order to indicate a language, in which the nature of all elements could be expressed

In this passage, Lorenz is becoming a bird, more precisely a crossbill or *Loxia*, i.e. a bird belonging to the finch family, whose main features are the mandibles with crossed tips. This detail is an evident reference to Lorenz's crisis of language, as, at this time, he can only communicate with inanimate objects, such as the mirror. For this reason, he appears like a human being, who has lost the principle human attribute, the speech. Because of aphasia and the following anthropological unsuitability, Lorenz is condemned to commit suicide, as he can no longer manage this precarious psychological and emotional condition. The text explains that he cannot find that life force in his soul, a force possessed instead, for example, by the tiger while jumping through the fire circle. Consequently, Lorenz can just embody the image of a meek roe deer, which, in this case, is torn not by hounds, but by the hidden, though deep pain due to his own male impotence:

> LORENZ (*kommt vor den Spiegel gelaufen*) Kein Kauz! Kein Phantast! Kein scheues Reh! Kein Figernagelbeißer! Kein Grübler, kein Pedant, kein Nuschler will ich sein… Der Feuerreif, durch den der Tiger springt… Vor mir: der Feuerreif, durch den der Tiger springt. Scheuen, Fauchen, und dann: ein Satz – ein *herrlicher* Satz![7]

The impotence motif seems to suggest that in this act the strongest character is represented *ex negativo* by Delia, the main female character, as she overcomes Lorenz's metaphorical predation attempt, just as her mythical precursor Diana, whose name also alliterates with the name of the Straußean figure. If a kind of woman superiority characterizes *Lorenz vor dem Spiegel*, it should be observed that in the same play also women can be 'animalized'; therefore, both basic gender categories are being discussed in Botho Strauß's *Schlußchor* exactly through the process of animalization, which constitutes at any rate also an occasion to query self-definition as well as human identity.

In the third act of the play, *Von nun an* ('Henceforth'), the plot focuses on the female character Anita von Schastorf and her animal 'regression', or better metamorphosis, which again reveals a mythical foundation. The woman is the daughter of a monarchist, murdered by the Nazis

---

directly; at the same time, *Spiegelatur* is an alternative to the language use in everyday life. Cf. Franziska Schößler, *Augen-Blicke. Erinnerung, Zeit und Geschichte in Dramen der neunziger Jahre* (Tübingen: Narr, 2004), p. 149.

7   *Schlußchor*, p. 55.

in 1944, and now trustee of her father's writing. During this act, Anita shows to be quite confused about the quick socio-political changes in her country, as the setting evokes the fall of the Berlin Wall through precise historical references, such as the exceptional opening of the frontiers between East and Western Germany. For this reason, Anita's anthropological disorientation can first be connected with this important political subject, which strongly influenced not only the socio-political orientation of many German as well as European citizens, but also their intimate psychological definition.[8] In this context, Anita's experience at the end of the act should be especially stressed, because it is set in the Berliner Zoological Garden, where her main aim is to visit the cage of the golden eagle, which also constitutes the German heraldic symbol. Again, in these pages, the leitmotif is exactly the pursuit of that 'naked truth', which emerges in the Derridean thoughts too: Anita praises this bird of prey because of its dignity and authenticity, which come from its 'nudity', as the bird is only covered with feathers, not with clothes. Unlike animals, humans cover themselves with garments and use these means to hide their psychological and social unsuitability. Therefore, they are always disadvantaged if compared to animals, since the latters do not need to cover themselves and do not feel shame; consequently, the fact of being naked stands for their metaphysical status of true and pure living beings:

> ANITA [...] (*Sie nimmt ein Messer aus ihrer Handtasche und schneidet ein großes Loch in den Draht der Voliere*) Wir sind ja unter unsern Kleidern sehr, sehr bloß. Davon ahnt ein Vogel nichts, wie arm das ist, so nackt und hilfsbedürftig, ohne Kleid. Damit hast du schon bei mir gewonnen, daß du nichts auszuziehen mußt. Daß du so schön bist, wie du aussiehst. Stolz in einem Stück. Nichts darunter, nichts dahinter. Federkleid vom Schädel bis zum Lauf. Mmmh: gute Lösung![9]

In the previous excerpt, Anita shows a kind of rooted envy towards the eagle and because of her feelings, she decides to offer the bird a prey, as sign of respect, also in order to convince the bird to mate with her. Through this detail, *Schlußchor* describes a very ambiguous situation, which also represents a sort of antithesis of the traditional concept of

---

8   Cf. Schößler, pp. 208-213. See Dag Kemser, *Zeitstücke zur deutschen Wiedervereinigung: Form – Inhalt – Wirkung* (Tübingen: Niemeyer, 2006), pp. 65-78, 145-146.

9   *Schlußchor*, p. 95.

sacrifice.[10] While in the primordial sacrifice scene, just like that of Abel, the individual gets in contact with God through the sacrificed animal victim, Anita breaks here a sacred taboo and yields in a Dionysian way to the Other-than-self.

In this context, Botho Strauß's consideration of the sacrifice theme is quite different from the corresponding representation in Derrida's text, because the German dramatist concentrates himself in this play especially on the political value of this over-species-mating, but implicitly also on the ritual and mythical significance of this orgiastic gesture. It should be highlighted that Anita's coupling with the eagle has been fulfilled after her failed attempt to change into a bird. It is here worth noting that this process of animalizing metamorphosis firstly occurs through a linguistic approach to the eagle: Anita imitates it by uttering inarticulate sounds, she cries out 'giuä, giuä',[11] and shows in this way her will to be no longer considered as a human being, endowed with the *logos*. Secondly, she imitates the animal physically, in particular through 'animal gestures', as she tries to ruffle up her clothes, as if she were a bird too:

> Ja, er schmiegte sich an mich, er setzte mir die Fänge auf die Hüfte und gellte: 'Anita [...] nun sträub dich nicht... laß uns vergessen... steig mit mir auf zum Himmelsrand!' So gellte er, ergriff mich schon und wollte sich erheben, doch ich wehrte mich. Seine Flügel schlugen in mein Kleid. Ich sträubte mich, ich sträubte mich.[12]

In this case Anita fails in her attempt to overcome the frontiers between animality and humanity, though in opposition to the second act of *Schlußchor*: Anita realizes that she is not able to reach the pure nakedness of the eagle; because of her garments, from which she cannot free herself, she ends up representing a lower living being. This result is illustrated through a semantic decline, i.e. the protagonist speaks no more of clothes, but instead of laundry: 'Ich bin Wäsche, Wäsche durch und durch, überall Wäsche. Unmöglich, mich bis auf meine Haut zu treffen' ('I am dirty laundry, throughout laundry, all over laundry. It is impossible to hit me up to my skin').[13] Therefore, the text suggests that

---

10   Cf. Alexis Joachimides and others (ed. by), *Opfer – Beute – Hauptgericht: Tiertötungen im interdisziplinären Diskurs* (Bielefeld: Transcript Verlag, 2016), pp. 9-132.

11   *Schlußchor*, p. 96.

12   Ibid.

13   Ibid., p. 98.

between the purity of feathers and Anita's 'cloths' can only exist an unbridgeable gap, which is precisely too wide to be closed because the animal represents metaphysical truth and the woman, on the contrary, a 'soiled' reality, i.e. untruth. Due to this impasse, this act of *Schlußchor* ends up again with the death motif, although this time it is not a human who dies, but an animal: the golden eagle is torn from a furious and deceived woman, who now plays the role of a new maenad. That is the reason why Botho Strauß's drama closes with the violent image of Anita's bleeding face; moreover, she holds a ripped talon and leaves the scene shouting for four consecutive times 'Wald' ('wood'), the absolute legendary place of the whole German-speaking tradition and in this play the symbol of a lost Germany.

The unsuccessful copulation between the woman and the bird is another reference to a Dionysian foundation as well as to the well-known myth of Zeus and Europa: a virgin Phoenician princess, who had been abducted by the God in the guise of a white bull and taken to Crete. On this island, Europa is seduced by Zeus, interestingly in the guise of an eagle, and gives birth to Minos. Considering this mythical reference, it is particularly important to underline the opposite process described in *Von nun an*. This act not only reverses the relationship between abductor and abductee, but it also presents a new issue within the domain of Animal Studies, which is related again to the problem of the political and power relations between species. In this context, Botho Strauß touches on a very awkward subject, i.e. zoophilia, a particular sexual attitude, which can be defined as 'a paraphilia characterized by recurrent sexually arousing fantasies, sexual urges, or behaviour involving erotic attraction of a person to an animal'.[14] For this reason, the spectators of *Schlußchor* are forced to experience – through the *pièce* – a mixture of sexual perversion and violent drive, which is to interpret as the natural consequence of the general atmosphere of brutality, fury and irrationalism of this work.

The end of *Schlußchor* describes in-between human-animal manners, mirroring the beginning of the *pièce*, as Anita's desperate quest for sense and socio-political recognition in the Berliner Zoo represents a variation

---

14   'Zoophilia', in *A Dictionary of Psychology*, ed. by Andrew M. Colman, 3rd edn, (New York: Oxford University Press, 2009), p. 831. Cf. Hani Miletski, 'Zoophilia: Another Sexual Orientation?', *Archives of Sexual Behaviour*, 46, Issue 1 (January 2017), 39-42. See Vv.Aa., *Human-Animal Studies: Über die gesellschaftliche Natur von Mensch-Tier-Verhältnissen*, ed. by Chimaira – Arbeitskreis für Human-Animal-Studies (Bielefeld: Transcript Verlag 2011), pp. 215-278.

of a Dionysian frame, which gives unity to Strauß's postdramatic work. Therefore, also the first act of the play, *Sehen und Gesehenwerden* ('To see and to be seen') – which is thematically connected to the sight motif of the second act and above all to Derrida's conference – includes several patterns of human-animal-violence as well as the topic of predation. This act ends up with a double representation of the Bacchic ritual named σπαραγμός (*'sparagmos'*), which consists in the dismemberment of a victim, indifferently human or animal, by a group of Bacchantes during a Dionysian orgy.[15] At first glance, this rite seems to be celebrated in *Sehen und Gesehenwerden* among members of a same biological category, i.e. mankind, although here different roles can be identified: the characters acting in these scenes are a choir, that is a social self-governing group – exactly like a herd – a photographer, and a passer-by. The first *sparagmos* takes place between the singers and the photographer; the second one between the musicians and an unfortunate woman. As in the second and third act of the play, also in this first act the dramatist builds the plot making reference to a mythical substrate. In this case, the recalled legend is that of the king of Thebes, Pentheus. He tried to interrupt a Dionysian celebration on the Mount Cithaeron, to which many women of Thebes, including his mother Agave and his two aunts Ino and Autonoë, were participating. Because of this affront, Dionysus wanted to revenge and obtained Pentheus's death through a savage murder: as the Bacchantes discovered the king behind a tree, they suddenly believed him to be a wild animal; therefore, they captured him and tore him limb from limb.[16] In the same way, the choir of the Straußean play is blind with orgiastic frenzy and its fifteen members agree to tear firstly their photographer, secondly the unknown woman.

Even though all these characters are human, it should be observed that in this passage the text is conceiving a strongly 'animalized' background. This aspect emerges from the lexical analysis of the act

---

15   The theme of the *sparagmos* is often revisited in postcolonial literature, as also Justine McConnell, lecturer in Comparative Literature at the King's College, recently pointed out. Cf. Justine McConnell, 'Postcolonial Sparagmos: Toni Morrison's *Sula* and Wole Soyinka's *The Bacchae of Euripides: A Communion Rite*', *Classical Receptions Journal*, 8, Issue 2 (2016), pp. 133-154.

16   Cf. 'Pentheus', in *The Oxford Classical Dictionary*, ed. by Simon Hornblower, Antony Spawforth and Esther Eidinow, 4th edn (Oxford: Oxford University Press, 2012), http://www.oxfordreference.com/view/10.1093/acref/9780199545568.001.0001/acref-9780199545568-e-4851 (22.03.17).

as well as from the accentuation of all its animal references; because of them, the boundaries between the human and the animal identity of the singers are constantly deconstructed. Starting from the choir – which can appear in the text also under the definition of *Meute* ('pack') and *Schar* ('flock') – the reader can notice a progressive utilisation of terms or nouns related to the animal domain; moreover, this act is full of idioms and expressions, which refer to an animal context, or, again, it presents a wide variety of substantives, which, on the level of the signifier, are formed by morphemes that derive from animal substantives.

The morphological extension which develops from this rich semantic field ranges from reptiles to mammals, insects and birds, and is composed by the following expressions, in which several entries of zoological origin can be recognized: *ins Auge des Reptils* ('in the eye of the reptile');[17] *Keiner verläßt die Wabe* ('nobody leaves the honeycomb');[18] *Immer mit den Wölfen heulen, bloß ein bißchen länger als das Rudel* ('always howl with wolves, but just a little bit longer as the pack') and *Reitsport* ('horse-riding');[19] *mit acht silbergrauen Huskies* ('with eight silver-grey huskies') and *was Vögel können im Flug* ('what birds can do while flying');[20] *Warum [...] gibst du das Biest nicht frei?* ('why don't you release the beast?');[21] *Sie Entensterz!* ('You, duck-rump!').[22] In this context, the author resorts to different ways of saying and wordplays as well as to epithets, which are used as terms of abuse. In addition, a wide range of idiomatic phrases, such as *in der Schlinge stecken* ('to get stuck in a trap'),[23] can be added to this list, although this act is also enriched which a particular range of expressions, which evoke a fantastic dimension or rather a Gothic atmosphere: in this case, the text creates a particular bloody and terrifying mood, as this act often refers to terms like *Ungeheuer* ('monster'), *blaurote Narbe* ('red-blue scar'), *geblutet* ('bled') or *Mordversuch* ('assassination attempt').[24]

If *Sehen und Gesehenwerden* is particularly rich in the use of an 'animal vocabulary', it should be stressed that this linguistic feature is an important element which permits to unify the whole structure of the

---

17 *Schlußchor*, p. 9.
18 Ibid., p. 13.
19 Ibid., p. 18.
20 Ibid., p. 20.
21 Ibid., p. 22.
22 Ibid., p. 25.
23 Ibid., p. 32.
24 Cf. ibid., pp. 10, 21.

drama *Schlußchor*. In fact, also in *Lorenz vor dem Spiegel* a widespread utilization of animal metaphors and comparisons can be found. For example, within the description of the party in the beautiful villa, the text refers to some guests through 'animal' epithets, which are also used as surnames: a man's name is for example 'Ferdinand Murrvogel' and his particular family name literally means 'grumbling bird'. Furthermore, the character called *Der Rufer* ('the person calling') continuously boasts to be acquainted with every kind of woman, although no woman he knew could be considered a bad person. This information is obviously expressed with a very ironic tone and is combined with the interesting fact that *Der Rufer* tells other guests about his relationships and lovers referring to animal nicknames, such as *Schmetterling* ('butterfly'), *Spatz* ('sparrow'), *Löwin* ('lioness') and even *Leseratte* ('bookworm'), which literally means 'reading rat'.[25]

Regarding this lexical constellation, it should be now questioned the nature of the boundaries between the 'anthropological machine', as the Italian philosopher Giorgio Agamben defines it, and 'animality' as well as the functioning of that 'open' space, or rather of that 'open' frontier between human and animal beings in Botho Strauß's works.[26] In fact, thanks to the wide utilization of an 'animal language' as well as of its animal rhetoric, the play *Schlußchor* can be qualified as an 'animal text' and functions as an appropriate literary example of the Zoological Aesthetics.[27] This fundamental concept within the field of Animal Studies can be firstly applied to the understanding of the representation of animals in several art forms. Secondly, Zoological Aesthetics stands for the way in which animals become 'actors', i.e. subjects of an autonomous aesthetics; and thirdly, this research horizon bases on a theory of animality, through which fundamental aesthetic topics are discussed, in particular from an 'animal point of view', or rather an 'animal perspective'.[28] According to these assumptions, animals do not simply play the role of 'object' of art representation, but they also participate actively in the construction and the definition of relevant

---

25   Cf. ibid., pp. 51-52.
26   Cf. Giorgio Agamben, *L'aperto. L'uomo e l'animale* (Torino: Bollati Boringhieri, 2014), pp. 38-43, 60-65.
27   Cf. Rodolfo Piskorski, *Animal as Text, Text as Animal*, in Reingard Spannring and others (ed. by), *Tiere, Texte, Transformationen. Kritische Perspektiven der Human-Animal Studies* (Bielefeld: Transcript Verlag, 2015), pp. 245-262.
28   Cf. Roland Borgards, Esther Köhring, Alexander Kling (ed. by), *Texte zur Tiertheorie* (Stuttgart: Reclam, 2015), pp. 18-21.

theoretical and aesthetic questions; moreover, their presence is always to be interpreted as an important occasion for poetic (self-)reflection, especially about the principal figures of speech, such as metaphors and analogies.[29] For this reason, the analysis of *Schlußchor* functions as a literary explanation of Cary Wolfe's famous declaration 'the animal, when you think about it, is everywhere',[30] because in this *pièce* human behaviour is strongly determined by the confrontation between humans and their deepest 'animal origins'.

Humans as Monstrous Beings: Der junge Mann

All previous poetic features deeply mark Botho Strauß's theatrical production, even if it can be demonstrated that they also characterize his earlier works, especially his prose. Themes like mythology, violence, brutality and animality as well as the exploration of anthropological frontiers are elements which also define the early novel *Der junge Mann* (1984).[31] Exactly as the postdramatic play *Schlußchor*, this text presents various internal sections, which immediately convey the impression of fragmentation, incoherence and disjointedness, and yet they still recall each other thanks to leitmotifs or thematic echoes, in other words thanks to a delicate thread running through the whole structure.

From the point of view of the Animal Studies, the analysis of the second chapter of this novel is particularly relevant. First of all, the chapter title is *Der Wald* and, therefore, it is dedicated to the legendary place and setting of all the best-known German fairy tales, the wood. Nevertheless, this chapter is far from telling a fairy tale as well as its protagonists are far from being heroes in a traditional sense. The figures playing in *Der Wald* are a hunter, a rude and dictatorial man, who is described as 'ein gemeiner Bonze' ('a mean bigwig'),[32] and a recluse prostitute, who is also his tenant, even though she detests him: 'Mit ihm hat es noch eine grausliche Bewandtnis' ('With him it is still a horrible situation').[33]

---

29   Cf. John Berger, *Why Look at Animals?*, in John Berger, *About Looking: Writers and Readers* (New York: Vintage International, 1980).

30   Cary Wolfe, 'Human, All Too Human: 'Animal Studies' and the Humanities', *PMLA*, Vol. 124, No. 2 (Mar., 2009), pp. 564-575 (p. 564).

31   See Botho Strauß, *The Young Man*, trans. by Roslyn Theobald (Evanston, Ill.: Hydra Books, 1995).

32   Botho Strauß, *Der junge Mann* (München: dtv, 2003), p. 71.

33   Ibid.

The characterization of the hunter is quite ambiguous, because he seems to live a kind of split personality: during the day, he is 'ein harmloser Wattefabrikant' ('a harmless cotton manufacturer'), but in his deepest soul he is above all a hermit hunter, withdrawn in his own house. This definition, i.e. *Der zurück in sein Haus gestopfte Jäger* ('the hunter, who is pushed back in his home'), is presented in the text as an independent headline, as if it would introduce a brief chapter within this section. Through this second title, which is thematically strictly connected to the semantic field of the wood, the novel points out that the hunter lives in hard social conditions, since he finds himself in a sort of imprisonment, or rather of confinement. The most relevant element in this context is the fact that the hunter is condemned to isolation from a public institution, the police. This is made clear in the chapter thanks to the narration of the pursuit of some bank robbers on the part of the police. In particular, the police are searching for a hidden haul in the wood and forbid everyone to interfere with the investigation, that also means that under these special conditions, the hunter is unable to practise his intimate passion, hunting.

In this context, it should be remarked that the 'imprisonment' of the figure causes himself an uncontrollable breaking out of bestial drives, because captivity makes the hunter mean and nasty – by Romance languages, the etymologic affinity between 'captivity' and 'bad' is clearer, as both substantive and adjective have as common linguistic root the Latin word *captivus*, which originally meant 'war prisoner' and came later to the meaning of 'wicked'. Very interesting in text is also the gradual transformation of the character into an animal. This metamorphosis begins as 'eine leise Verstimmung'[34] ('a light bad mood') and reminds, in its structure, the famous metamorphosis, which constitutes the main theme of Franz Kafka's novella *Die Verwandlung* ('The Metamorphosis', 1915). For this reason, it also ends up representing one of the narrative peaks of Botho Strauß's whole production, in particular, in relation to the metaphysical as well as psycho-biological 'confusion', or rather in relation to that 'melting point', which characterizes the fluid separation of humanity and animality.

Exactly as in *Schlußchor*, the hunter's transformation occurs within a Dionysian context, which is signalled in these pages through the presence of a distorted erotic atmosphere. This almost grotesque background is summarized in the dialogue between an accountant and the bizarre prostitute, who lives partially in the wood and partially in

---

34   Ibid.

the hunter's hide. The Bacchantic atmosphere gets stronger thanks to a particular rhetoric device, the climax, through which the hunter's ill humour comes to a definition. As a matter of fact, the more the figure is forced to wait inside the house, the more his bad feeling becomes a blind rage and later even a bloody foolishness:

> Gewaltsam daran gehindert, sein Haus zu verlassen, seine Beute zu machen, verwandelte sich ihm zuerst das Hobby, der Ausgleichssport, der Wille zur Freizeit in eine angestaute Leidenschaft. Sodann wandelte sich der Leidenschaftsstau in ein innerliches, aber doch schon schweres Verbrechertum, und dieses schließlich, ebenfalls unterdrückt, wandelte sich in eine nur kurz vorm Ausbruch zurückgehaltene Bestialität. Über diese Stufen der Gierverwandlung war der Jäger von einer leisen Verstimmung bis bald an die Grenze seiner Entartung vorangeschritten. Übermäßige Pressung und Drosselung eines harmlosen Jagdeifers hatten dazu geführt, daß eine gefährliche Willenskernverschmelzung in ihm stattfand, eine gewaltige Energieverdichtung, die ein Mensch kaum mehr als er selber zu überleben vermag. In Wahrheit gab es nun für ihn kein Halten mehr, er hatte ja längst den Gipfel seiner Verformung überschritten, er *war* bereits der nun noch spärlich verkleidete Wolfsmensch oder Blutsauger, der nun freilich nicht mehr auf Jagd nach Niederwild auszog.[35]

As this passage shows, the hunter's transformation into a predator and a monster is certainly gradual, but again it is not possible to outline accurately the moment in which his animal degeneration transcends humanity and falls into animality. When it is all over, the narrator simply inquires: 'Was aber, da er die äußerste Landspitze des Menschseins erreicht hatte, sollte nun weiter mit ihm geschehen?' ('What else should it happen then with him, since he reached the extreme headland of personhood?').[36] Consequently, the bestial metamorphosis of the hunter makes the readers understand what 'humanity' is not, or what it is no more, through a process *ex negativo*: the text describes the hunter's terrible transformation into a violent living being, but it does not explain in what humanity (or animality!) definitely consists. Therefore, the hunter loses his biological identity of human being precisely in the moment in which he gives himself up to an extreme survival impetus, which is strictly linked to a long time suppressed instinct and so to

---

35   Ibid., p. 72.
36   Ibid.

an inhibition process too. Because of this factor, the hunter becomes a vague entity, in between a werewolf and a vampire.

Compared to the psychological and social definition of the individual, this withdrawal is quite relevant and it acquires a significant position also in relation to the chapter's structure, as the hunter's metamorphosis concludes the wood episode in *Der junge Mann*. In particular, the last scene of the section *Der Wald* describes how the hunter welcomes the police in his home and tries, though with difficulty, to restore his middle-class habit, in German 'seine brüchige äußere Bürgerhülle' ('his fragile external middle-class-cover').[37] By this gesture, he also takes advantage of the opportunity to scan the horizon from the wide-open door, in order to establish again a contact with his old passion as well as with the hunting environment.

However, it should be highlighted here that because of his internment, the hunter's interest in hunting as well as in the wood has decisively decreased; consequently, the wood now represents for the character just a lifeless place, 'ein trübes, lebloses, verschwommenes Feld' ('a bleak, lifeless, blurred field').[38] On the contrary, in this final moment the hunter identifies a new goal and so new victims too: he does not want to shoot animals anymore, because he now perceives them as fellow creatures; instead, he considers humans as his new preys, because they do not represent his own kind anymore. In this way, the character shows to be aware of his new ambiguous identity and tries to establish a new predatory relation looking at the policemen as legitimate sacrificial victims: 'Das Pirschraster seiner vergrausten Augen erfaßt nun den Rücken des letzten, ein wenig hintangebliebenen, blutjungen Beamten…' ('the retina of his savage eyes now seizes the back of the last and very young police officer, who remained a little bit backwards…').[39]

In this last sentence, the text refers again to the central sight-discourse and defines it with a new 'view', i.e. through the look of a shapeless living being, which, or rather who, is no more human and not completely animal. Moreover, this metaphysical vague status is confirmed in the text from the three dots, and so from the graphical representation of something which cannot be explained from the narrator, i.e. it cannot be expressed through the human *logos*. What is more, these dots predict a new *sparagmos* ritual, as the hunter is killing

---

37   Ibid.
38   Ibid., p. 73.
39   Ibid.

a man by virtue of his Dionysian (re-)awakening and exaltation. Due to this Bacchic excitement, the hunter loses the Apollonian control on his own person and transforms into a new unknown creature, actually a living being, through which Botho Strauß definitively questions what the real boundaries between anthropology and zoology are.

Mythology, Animality, and the Sense of Sight: Concluding Considerations

In conclusion, this essay aims to illustrate how myth and modernity are combined and melt together in Botho Strauß's works as well as why this relation becomes significant compared to the research field of Animal Studies. As Sigrid Berka points out, this important connection can never be cleared once and for all, especially on the dialectical level. On the contrary, this relationship constantly bases on a 'fluid' confrontation, but also on an irreconcilable difference.[40]

Interestingly enough, this particular relation has also to be seen within the illuminating parallel between human and animal figures by Botho Strauß; at the same time, it concerns the analogies between humans and other nebulous characters as well as between human beings and complex heroes, who challenge their own human identity.

However, the insisted postponement of this psychological border always forces a careful problematization of the human nature by Botho Strauß: although the hunter becomes a beast, his animalization's process is not described directly, but from an external point of view, i.e. from the narrator's perspective, which is anything but neutral and so it widely influences the reader's reception or interpretation of the text. Actually, we can assume that this narrator is characterized by a bourgeois moral and therefore, he simply cannot be identified with the author himself. For this reason, the hunter loses his bio-psychophysical equilibrium only in the opinion of this invisible narrative instance; on closer inspection, it is still the hunter, who, in the final scene, 'seizes' with his sight the belated policeman. And his look is, following Derrida,

---

40 'Laut Berka zeigt das Spiel mit den Mythen darüber hinaus, daß Mythos und Moderne nicht dialektisch vermittelbar sind, sondern im Verhältnis unauflöslicher Konfrontation stehen'. Cit. in Herbert Grieshop, *Rhetorik des Augenblicks: Studien zu Thomas Bernhard, Heiner Müller, Peter Handke und Botho Strauß* (Würzburg: Königshausen & Neumann, 1998), p. 207.

still the glance of the individual who says *je suis*, that is 'I am' as well as 'I follow': even if this look may be guilty, it still represents a human gesture, a gesture which marks a very human action, beyond every kind of rigid psychological, political and social construct.

Bibliography

Primary literature

Agamben, Giorgio, *L'aperto. L'uomo e l'animale* (Torino: Bollati Boringhieri, 2014).
Derrida, Jacques, and David Wills, 'The Animal That Therefore I Am (More to follow)', *Critical Inquiry*, 28, No. 2 (Winter, 2002), 369-418.
Strauß, Botho, *Der junge Mann* (München: dtv, 2003).
Strauß, Botho, *Schlußchor* (München: dtv, 2007).
Strauß, Botho, *The Young Man*, trans. by Roslyn Theobald (Evanston, Ill.: Hydra Books, 1995).

Secondary literature

Berger, John, *About Looking: Writers and Readers* (New York: Vintage International, 1980).
Berka, Sigrid, *Myhthos-Theorie und Allegorik bei Botho Strauß* (Wien: Passagen-Verl., 1991).
Borgards, Roland, Esther Köhring and Alexander Kling (ed. by), *Texte zur Tiertheorie* (Stuttgart: Reclam, 2015).
Colman, Andrew M., *A Dictionary of Psychology*, 3rd edn (New York: Oxford University Press, 2009).
Dalmasso, Gianfranco, 'Introduzione all'edizione italiana: Il limite della vita', in Jaques Derrida, *L'animale che dunque sono*, ed. by Marie-Louise Mallet, and Gianfranco Dalmasso (Milano: Jaca Book, 2006), pp. 7-28.
Grieshop, Herbert, *Rhetorik des Augenblicks: Studien zu Thomas Bernhard, Heiner Müller, Peter Handke und Botho Strauß* (Würzburg: Königshausen & Neumann, 1998).
Huller, Eva C., *Griechisches Theater in Deutschland: Mythos und Tragödie bei Heiner Müller und Botho Strauß* (Köln u.a.: Böhlau, 2007).
Joachimides, Alexis and others (ed. by), *Opfer – Beute – Hauptgericht: Tiertötungen im interdisziplinären Diskurs* (Bielefeld: Transcript Verlag, 2016).
Kemser, Dag, *Zeitstücke zur deutschen Wiedervereinigung: Form – Inhalt – Wirkung* (Tübingen: Niemeyer, 2006).
McConnell, Justine, 'Postcolonial Sparagmos: Toni Morrison's *Sula* and Wole Soyinka's *The Bacchae of Euripides: A Communion Rite*', *Classical Receptions Journal*, 8, Issue 2 (2016), 133-154.
Miletski, Hani, 'Zoophilia: Another Sexual Orientation?', *Archives of Sexual*

Behaviour, 46, Issue 1 (January 2017), 39-42.

Piskorski, Rodolfo, *Animal as Text, Text as Animal*, in Reingard Spannring and others (ed. by), *Tiere, Texte, Transformationen. Kritische Perspektiven der Human-Animal Studies* (Bielefeld: Transcript Verlag, 2015), pp. 245-262.

Schößler, Franziska, *Augen-Blicke. Erinnerung, Zeit und Geschichte in Dramen der neunziger Jahre* (Tübingen: Narr, 2004).

Vv.Aa., *Human-Animal Studies: Über die gesellschaftliche Natur von Mensch-Tier-Verhältnissen*, ed. by Chimaira – Arbeitskreis für Human-Animal-Studies (Bielefeld: Transcript Verlag 2011).

Vv.Aa., *The Oxford Classical Dictionary*, ed. by Simon Hornblower, Antony Spawforth and Esther Eidinow, 4th edn (Oxford: Oxford University Press, 2012).

Wolfe, Cary, 'Human, All Too Human: 'Animal Studies' and the Humanities', *PMLA*, Vol. 124, No. 2 (Mar., 2009), 564-575.

JUDITH RAHN

HUMAN BEASTS: EATING MEAT AS A NEGOTIATION OF SELF

Relationality: Between the Animal and the Human

The controversial relationship between humans and animals has been fundamentally negotiated, questioned, and re-negotiated since the beginning of humankind. The necessity for a thorough consideration of human-animal boundaries is therefore at the core of human anthropocentric and speciesist thinking, while the question of edibility plays a major role in the negotiation of cultural and moral norms. Peter Singer among others[1] in the academic and recently Jonathan Foer[2] in the popular discourse, famously make a strong case for humankind's responsibility towards animals by calling especially for a consideration of human-animal similarities in the ability to suffer ('If a being suffers there can be no moral justification for refusing to take that suffering into consideration'[3]).

This paper will examine the correlation of animal meat consumption and anthropophagy as a necessary force in the attempted (re-) construction and renegotiation of the human, the animal and the edible. Man eating animal and man eating man regains a strange visibility, which reflects upon an almost freakish intimacy between the cannibal and the consumed subject, and increasingly blurs the boundaries between the animal and the human. However, recent posthumanist research[4] has not only questioned ethical and moral boundaries between what we understand to be human and what we consider to be

---

1 Peter Singer, *Animal Liberation*, 3rd edn (New York: Ecco, 2002 [1975]); See also: Tom Regan, *The Case for Animal Rights*, updated edn (Berkeley, London: University of California Press, 2004).

2 Jonathan Safran Foer, *Eating Animals*, (London: Hamish Hamilton, 2009).

3 Singer, p. 8.

4 Cf. e.g.: Cary Wolfe, *What Is Posthumanism?*, (Minneapolis: University of Minnesota Press, 2010); Rosi Braidotti, *The Posthuman*, (Cambridge: Polity, 2013); Katherine Hayles, *How We Became Posthuman: Virtual Bodies in*

animal, but also elicited the fundamental materiality and relationality between species, even questioning species boundaries altogether. However, there has been ample discussion regarding flexibility and necessity of differentiations between notions of anthropocentrism and speciesism.[5] The belief that the 'characteristics possessed by humans entail moral superiority'[6] is essential in the debate of animal and human subjectivities, and the idea of deconstruction of traditional subjectivities has recently entered academic discourse, as underlined by Spiegel: '[i]t is only an anthropocentric world view which makes the qualities possessed by humans to be those by which all other species are measured'.[7] Jacques Derrida, for example, has famously deconstructed epistemic assumptions discerning animals and humans by challenging the normativity of anthropocentric hierarchies, which are also inherent in allegedly non-anthropocentric thought. Although this view has been widely and favourably received,[8] it has also encountered equally widespread criticism, as for example Gary Steiner's, who – in questioning ideals of truth – argues: 'Derrida fails to articulate any clear moral principles bearing on our relationship to animals'[9] which he sees as vital in the improvement of human-animal relationships. As controversially contested as these conceptions may seem, they all confirm the need to rethink epistemic conventions and newly consider the positions of both humans and animals. These tendencies of breaking through the species boundary in contemporary theoretical and philosophical thought is, of course, equally reflected in (popular) culture. this is visible in the

---

Cybernetics, Literature, and Informatics, (Chicago; London: University of Chicago Press, 1999).

5   The idea of 'speciecism' as an analogy to 'racism' was first addressed by Richard Ryder: Richard Ryder, 'Experiments on Animals', in *Animals, Men and Morals: An Enquiry into the Maltreatment of Non-Humans,* ed. by Stanley Godlovitch, Roslind Godlovitch, and John Harris (London: Gollancz, 1971), pp. 41-82 (p. 82).

6   Andreas Aigner, Karl Pieper, and Herwig Grimm, "Post-Anthropocentrism' in Animal Philosophy and Ethics: The Disparity of the Prefix 'Post", *HUMaNIMALIA,* 7 (2016).

7   Marjorie Spiegel, *The Dreaded Comparison: Human and Animal Slavery,* Rev. and expanded edn (New York: Mirror Books, 1996), p. 23.

8   Cf. e.g.: Leonard Lawlor, *This Is Not Sufficient: An Essay on Animality and Human Nature in Derrida,* (New York; Chichester: Columbia University Press, 2007).

9   Gary Steiner, *Anthropocentrism and Its Discontents: The Moral Status of Animals in the History of Western Philosophy,* (Pittsburgh: University of Pittsburgh Press, 2005), p. 5.

interest in protagonists that are for example mythological half-human or half-animal hybrid creatures iterating the question not only of human singularity but also of the precarity of life – eventually resulting in essentially nonsensical creations like the vegetarian vampire.

If we consider the animal alongside the human, the *idea* of the cannibal becomes far more important than its factual realization: it is the kinship with the animal, as Animal Studies suggest, that creates these unsolved tensions between likeness and otherness, between food and kin, between human and animal. Before resolving the question of likeness between animals and humans, we must first resolve the question of edibility of animal (and also that of humankind). The problem here is the paradoxical relation between the moral problem of eating animals that can be considered akin to us – thus making humans effectively cannibals – or to reject our notion of kinship with animals altogether in favour of an unproblematic consumption.

Inter-Species Relations

Donna Haraway, in her 2008 publication *When Species Meet*, points out the long-standing engagement of mankind with human exceptionalism while confirming that it is, indeed a 'fantasy', which is 'the premise that humanity alone is not a spatial and temporal web of interspecies dependencies'.[10] The shift from human exceptionalism towards a broader, more interrelated approach has been fuelled by the need to re-think previously static parameters of normativity. Binary associations such as male-female, man-machine, human-animal are supported by a system of normativities, that were already challenged with the emergence of psychoanalysis. In fact, Haraway examines the impact Sigmund Freud has had on Western academic discourse, pointing out that:

> Freud described three great historical wounds to the primary narcissism of the self-centered human subject, who tries to hold panic at bay by the fantasy of human exceptionalism. First is the Copernican wound that removed Earth itself, man's home world, from the center of the cosmos and indeed paved the way for that cosmos to burst open into a universe

---

10   Donna Jeanne Haraway, *When Species Meet*, (Minneapolis: University of Minnesota Press, 2008), p. 11.

of inhumane, nonteleological times and spaces. Science made that decentering cut. The second wound is the Darwinian, which put Homo sapiens firmly in the world of other critters, all trying to make an earthly living and so evolving in relation to one another without the sureties of directional signposts that culminate in Man. Science inflicted that cruel cut too. The third wound is the Freudian, which posited an unconscious that undid the primacy of conscious processes, including the reason that comforted Man with his unique excellence, with dire consequences for teleology once again. Science seems to hold that blade too.[11]

The third wound highlighted by Donna Haraway addresses the necessity to re-think both hierarchical structures concerning the congruent relationship of body and mind and the impact this same hierarchical structure has on other binary oppositions. While the unconscious is, historically speaking, often seen as the counterpart of reason[12] it also functions to uphold the systems of hierarchy that govern the structures of body, mind, and reason. The 'primacy of conscious processes', that 'comforted Man with his unique excellence' is at the heart not only of the man-technology divide, but also at the centre of the connectivity between man and animal. Psychoanalysis is therefore often seen as an early extension of, among others, new materialist and also posthumanist thought[13] as it challenges the dominance of reason over body and mind, and rather postulates the unconscious as equalizing this binary bond.

While for Sigmund Freud the 'analyst is, among other things, an explorer of the dark continent of the unconscious and the repressed',[14] he has much in common with the first explorers of the colonial world, who discovered civilizations that used cannibalism as a standard cultural practice. Indeed, although 'psychoanalysis has focussed generally on the taboo of incest, Freud was equally interested in cannibalism, which, in *Totem and Taboo,* is posited as the dark secret behind the origins of civilization.'[15] This notion alone sees cannibalism in-between civilized

---

11 Haraway, p. 11.
12 This concept has repeatedly been discussed with regard to its political implications as it questions the ultimate rationality of free will. Cf. Sigmund Freud, *Die Traumdeutung*, (Leipzig; Wien, 1900).
13 Cf. Anthony Miccoli, *Posthuman Suffering and the Technological Embrace*, (Lanham, MD: Lexington Books, 2010), p. 83.
14 Francis Barker, Peter Hulme, and Margaret Iversen, *Cannibalism and the Colonial World*, (Cambridge: Cambridge University Press, 1998), p. 249.
15 Ibid.

and uncivilized, between notions of culture and ideas of animality. The history of Animal Studies has shown, however, that this simple divide is ridden with moral, ethical, and theoretical complexities which are mirrored in the multifariousness with which the trope of the cannibal is negotiated in contemporary theory and fiction.

What is the Animal?

Animal-human relations oscillate between the fictional and the factual. Among Animal Studies' theories and practices, the figure of the cannibal takes on a particular role of in-betweenness, a marginalized phenomenon. In the depiction of the cannibal, we experience a merging of human-nature divide similar to the act of cannibalism which reduces human flesh to animal meat, erecting and undermining seemingly clear-cut boundaries of the human. Conversely, the danger of cannibalism lies also in the implicit possibility for every human to become a cannibal as well as to become the victim of cannibalism. Cannibalism is never purely an act of cruelty because it marginalises the consumer as well as the consumed.

The degree of interaction between the eater and the consumed that transgresses the boundaries of death is unique to the act of cannibalism. Cannibalism portrays the ambiguous role of the consumed but also of the consumer, as it is outwardly an exertion of violence, power, and normativity which, at the same time, factually negotiates the hierarchical condition of the relationship between eater and eaten. Thus, the act of incorporation has transferred the unity of the body from the realm of certain definition into the unreliable dimension of ambivalence. All the while, the consumption of human flesh draws attention to the unutterable otherness of the eaten body but equally confirming its sameness, and its fitting similarity to the eater's body. Indeed, Matthew Calarco, in his publication *Zoographies: The Question of the Animal from Heidegger to Derrida* (2008), argues with regard to the status of the other that

> [...] violence is irreducible in our relations with the Other [...]. In order to speak and think about or relate to the Other, the Other must – to some extent – be appropriated and violated, even if only symbolically.[16]

---

16   Matthew Calarco, *Zoographies: The Question of the Animal from Heidegger to Derrida*, (New York: Columbia University Press, 2008), p. 136.

Here, Calarco draws a connection between self and other, the eater and
the eaten, the act of othering, and the thereby inherent appropriation. The
relationship between self and other can never be regarded as non-violent,
for the moment of sheer contact already includes the violent interference
with the other's integrity 'even if only symbolically'. This reference to
the symbolic has not to be easily dismissed, and plays as an important
role in the analysis of acts of cannibalism in fiction as it does in the study
of factual accounts of anthropophagy. Hence, Calarco continues:

> How does one respect the singularity of the Other without betraying
> that alterity? Any act of identification, naming, or relation is a betrayal
> of and a violence toward the Other. Of course, this should not be taken
> to mean that such violence is immoral or that all forms of violence are
> equivalent. Rather, the aim is to undercut completely the possibility
> of achieving good conscience in regard to questions of nonviolence
> toward the Other.[17]

By association, therefore, any act or relation towards the non-self, the
other, be it for example the edible or the animalic, contains the spark
of violation – which is not reprehensible per se, but argues towards the
impossibility of 'nonviolence' towards the other. As soon as the other
is identified as the non-self, it is appropriated, and therefore cannot be
regarded as untouched. By proposing to not believe in 'the existence of
the non-carnivore in general'[18] Derrida sees mankind's gaze towards this
conceptualization of the animal as automatically carnivorous, stating that
'in a more or less refined, subtle form, a certain cannibalism remains
unsurpassable.'[19] The precarious status of the cannibal becomes more
visible when considered in combination with Animal Studies, as the
cannibal highlights the inadequacy of the categories 'human' and 'animal'.
At this cross-roads between consumption and kinship of animals, the
cannibalistic human position is worth further consideration.[20]

---

17   Ibid.
18   Jacques Derrida and Elisabeth Roudinesco, 'Violence against Animals', in
      For What Tomorrow: A Dialogue, (Stanford: Stanford University Press,
      2004), pp. 62-76 (p. 68).
19   Ibid. p. 67.
20   cf. e.g.: Analaía Villagra, 'Cannibalism, Consumption, and Kinship in Animal
      Studies', in *Making Animal Meaning,* ed. by Linda Kalof and Georgina M.
      Montgomery (Michigan: Michigan State University Press, 2011), pp. 45-56.

As Jacques Derrida asks 'What animal? The other',[21] and continues to analyse the paradoxical human awareness of the inevitably anthropocentric perspective of regarding animals, he situates the animal in the unfathomable realm of 'the other'. However, it is the crossing of these seemingly and self-evidently separate categories that evokes social and moral taboos, rendering the animal hybrid a strangely marginalized non-entity. In his controversially discussed text *Eating Animals* (2009), Jonathan Safran Foer bases the roots of his argument – just as Peter Singer did in *Animal Liberation* (1975/2002) – on the reference to the essential sameness of the body that is so particularly visible in cannibalism and the inter-species relation between man and dog:

> I wouldn't eat George [the dog], because she's mine. But why wouldn't I eat a dog I'd never met? Or move to the point, what justification might I have for sparing dogs but eating other animals?[22]

If dogs are considered inedible in a Western context, and pigs are not, what does construct species boundaries and how valid are they? Following the line of argumentation of the anthropologist Carlos Fausto which considers cannibalism to be 'any devouring (literal or symbolic) of the other in its (raw) condition',[23] cannibalism can be used as a much broader term with regard to the intersection of animality and humanity. Fausto further continues:

> Noncannibal consumption supposes a process of desubjectifying the prey, a process in which culinary fire plays a central part. In daily meals, the animal-as-subject must be absent for identification to occur between humans. Any subjective relation between human and animal must be blocked so that the latter's meat can provide the medium for commensals to produce themselves as humans and relatives.[24]

This suggests that only by removing animal subjectivity from the process of consumption, do we achieve human identification. I would propose, however, that in order to productively utilise this contradictory

---

21   Jacques Derrida, 'The Animal That Therefore I Am (More to Follow)', *Critical Inquiry*, 28 (2002), p. 372.

22   Jonathan Safran Foer, *Eating Animals*. (London: Hamish Hamilton, 2009), p. 17.

23   Carlos Fausto, 'Feasting on People: Eating Animals and Humans in Amazonia', *Current Anthropology*, 48 (2007), 504.

24   Ibid.

phenomenon in Animal Studies and in the reading of literature, we must re-consider these binary notions of identification and kinship. In order to seriously consider the role of the cannibal as a precursor of fluidity between perceived species boundaries, we need to understand the human, the animal, and all species boundaries in equally relational terms. Focussing on relational, rather than absolute, terms can help to negotiate the nature of the identification with animals, but it still does not describe the almost paradoxical relationship among the cannibal as an in-between figure of animalistic humanity, the human, and the animal. If cannibalism encourages a position of power, it also questions the relationality with the realm of animality, and calls for an investigative engagement with the transformative nature of cannibalism. Anthropophagy does not only transform the eaten body, it of course also transforms the eater's body, through the consumption of what is – and must necessarily be – considered other (if we assume that the consumption of the self is not an option). However, in consuming the other, the other itself is promoted to become a part of the self, that is one of the most intimate connections possible, as the other is incorporated into the eater's own body and thereby made 'self' as well. This does not retain the unsolvable paradox of cannibalism which can be described as a state of inbetweenness, where the other is the non-self but also has the potential to turn into a part of the self by means of incorporation. Cannibalism, therefore, encourages a strangely intimate affinity to what is eaten and through the process of consumption creates an unavoidable relation with the other. Cătălin Avramescu in her 2011 publication *An Intellectual History of Cannibalism* asks, 'Are cannibals human beings?',[25] which I would argue, needs to be followed by the question 'Is eating animals cannibalistic'?

In her recent *MaddAddam* Trilogy (2003-2013),[26] Margaret Atwood combines tropes of cannibalism in a complex post-apocalyptic scenario with the creation of a new, herbivorous, human race to re-negotiate and re-evaluate contemporary attitudes towards animal consumption. Questions of consumptions are here negotiated by means of synthetically breed human-animal hybrids, that are ironically created for the sole purpose

---

25   Cătălin Avramescu, *An Intellectual History of Cannibalism*, (Princeton: Princeton University Press, 2009), p. 85.
26   The trilogy is comprised of (in order of publication): Margaret Atwood, *Oryx and Crake*, (London: Bloomsbury, 2003); Margaret Atwood, *The Year of the Flood*, (London: Bloomsbury, 2009); Margaret Atwood, *Maddaddam*, (London: Bloomsbury, 2013).

of being consumed. Pig-creatures with human intellect, goat hybrids with human hair, chicken-plants that produce only chicken-breasts and a newly created human race that rejects all animal food in favour of plant-consumption is Margaret Atwood's ironic vision of the human future.

Dissolving Boundaries in Margaret Atwood's MaddAddam Trilogy

The formulation of the notion of species in the consideration of metaphors of physical incorporation is an inherent question. Maggie Kilgour, in her seminal work *From Communion to Cannibalism* (1990), comments on the importance of cannibalism to definitions of identity, either 'physical, textual, or social',[27] and Simon Estok claims that, for a more complex understanding of the issue of cannibalism, one would need to consider the relevance of the term 'species' for Kilgour's list.[28] Of course, this is an immanent question in the consideration of metaphors of physical incorporation, since the line between species has become considerably more blurred with the emergence of new academic discourses as can be seen for example in ecocritical and posthumanist thought. This is very openly visible in Margaret Atwood's *MaddAddam* trilogy which is set in a post-apocalyptical world where the human race has been mainly eradicated by a fatal virus secretly devised by the prodigy, who turned into a 'mad scientist', codenamed 'Crake', and replaced by a supposedly superior, genetically altered, and pseudo-indigenous species, the 'Crakers'.

The narratives around the protagonist Snowman switches back and forth between the post-apocalyptic scenario of a world in ruins and flashbacks from before the catastrophe. Snowman finds himself living on a tree near a group of seemingly primitive pseudo-humans, which he calls the Crakers. Flashbacks continuously reveal the story leading up the apocalyptic events that destroyed the known world. A genetically modified virus designed to eradicate the human race is spread by Crake, a promising young scientist who decides to take the fate of humanity in his own hands by creating a superior human race. With the completion of his genetic experiments which resulted in the creation of the Crakers,

---

27   Maggie Kilgour, *From Communion to Cannibalism: An Anatomy of Metaphors of Incorporation*, (Princeton; Oxford: Princeton University Press, 1990), p. 16.
28   Simon C. Estok, 'Cannibalism, Ecocriticism, and Portraying the Journey', *CLCWeb: Comparative Literature and Culture,* 14 (2012), p. 2.

Crake decides to kill the entire human race – including himself and his associates – to give the new humans the chance of a fresh start. These apocalyptic events are often referred to as the 'Waterless Flood' to describe the inescapable force of the virus that seems to be as an uncontrollable natural force as the flood that safely carried Noah and his arch into a new world. In an ironic turn of events, Crake's plan fails to destroy the entire human race, which results in a small group of human survivors trying to make a living alongside the Crakes. In the course of the events, it is revealed that Snowman was once called Jimmy, a boy closely connected to – if not responsible – for the events that ended the world as he knew it, since he, Crake and a girl named Oryx were the centre of the team that designed the deadly virus. While Crake was responsible for the scientific conception of the Crakers, Oryx took on the role of social worker-teacher-nurse and educates the childlike Crakers according to Crake's wishes. Having grown up in a world neatly divided into the privileged employees of big corporations and the non-privileged rest, Snowman in hindsight re-evaluates the chances and responsibilities of the human race and the role that the Crakers play in this new world.

The Crakers fill a unique, and rather absurd position in Atwood's world, as they are designed to be the best version of humankind that is possible, yet turn out to exist somewhere between curious, childlike naivety and the beginnings of human evolution. To minimalize the chances of a war for food and the exploitation of the planet through humankind, their maker created the Crakers to not only be vegetarian but ruminant, living off grass and leaves.

> What had been altered was nothing less than the ancient primate brain. Gone were its destructive features, the features responsible for the world's current illnesses. [...] Since they were neither hunters nor agriculturalists hungry for land, there was no territoriality [...]. They ate nothing but leaves and grass and roots and a berry or two; thus their foods were plentiful and always available.[29]

Crake's ambivalent position within the narrative is due to his dual position as creator. While on the one hand he is the destroyer of humankind, he is on the other hand the spiritual father and creator of the Crakers (Snowman therefore also calls the Crakers 'The Children

---

29   Margaret Atwood, *Oryx and Crake* (London: Bloomsbury, 2003), p. 305.

of Crake'). Before the world is destroyed, Oryx, the only girl in the triangle between Oryx, Crake, and Snowman, is assigned to teach the Crakers' about what to eat, where to live, and that animals are not edible for the Children of Crake. She is considered mother of all things living, and the thought of eating 'The Children of Oryx' seems abominable to the Crakers. This causes some conflict when Snowman and the other remaining humans hunt, kill, and consume animals to survive. Snowman tries to explain to the Crakers how the 'Children of Oryx' tend to eat each other:

> [W]e [the humans] sat around the fire and ate soup. [...] Yes, there was a bone in the soup. Yes, it was a smelly bone. I know you do not eat a smelly bone. But many of the Children of Oryx like to eat such bones. Bobkittens eat them, and rakunks, and pigoons, and liobams. They all eat smelly bones.[30]

Vegetarianism is such an innate part of this species that the concept of living of anything but an herbivorous diet seems utterly alien, revolting, and other to the Crakers – while in reverse, the insistence on the inedibility of meat and bone locates these new humans more in the realm of the herbivorous animal than of the human. There are certain parallels between early depictions of cannibalism in the colonial world[31] and this scene of commonly shared animal meat during a feast at the beach.

Toby, the narrator of the third book in the *MaddAddam* trilogy, has taken on the task of conveying to the Crakers a sense of history which is a humorous undertaking, as their child-like innocence and curiosity poses a great problem to all those who try to establish a connection with them. This hybrid human-animal species, so familiar and yet so alien, stands at the threshold of a new-old world, that is so different after the apocalypse but which, at the same time, seems not to have changed much at all. It makes room for a multitude of beings that cross the species divide on more than just one level. In this alien world, it is the pigoon, an utterly strange and yet familiar pig-like animal, bred for its corporeal resources, that possesses most human-like intellectual abilities. Snowman as a child witnessed his father breeding these

---

30   Atwood, p. 4.
31   Cf. Peter Hulme, 'Introduction: The Cannibal Scene', in *Cannibalism and the Colonial World,* ed. by Francis Barker, Peter Hulme, and Margaret Iversen (Cambridge: Cambridge University Press, 1998), pp. 1-38 (p. 2).

super-pigs in an attempt to facilitate human organ growth in animals for easy transplantation afterwards. Glancing into their stable at the laboratory, Snowman has the fierce notion of being stared at, and of being understood by these animals, which makes the consumption of pigoons a difficult choice for the protagonist:

> 'Pigoon pie again,' they would say. 'Pigoon pancakes, pigoon popcorn. Come on, Jimmy, eat up!' This would upset Jimmy; he was confused about who should be allowed to eat what. He didn't want to eat pigoon, because he thought of the pigoons as creatures much like himself. Neither he nor they had a lot of say in what was going on.[32]

The idea of meat-eating and its almost invisible borderline with cannibalism advances to be a central problem in Atwood's post-apocalyptic trilogy, where food and especially meat supplies are luxury items and substitute products are being released to a market which cannot provide unrestricted access to meat and dairy foods. Instead, capitalistically abusive corporations substitute animal protein for soy products, while gen labs develop a so-called animal 'splice' which resembles a chicken-plant growing only chicken breasts. The text raises questions of responsible farming and potentialities and limits of scientific advances: is a chicken breast grown from a chicken-plant still a chicken breast? Is the consumption of pigoon meat unethical if pigoons are as intelligent as humans? Is there even a difference between eating human flesh and consuming animal meat if animals have the same intellectual abilities and physical needs as humans? While these questions are not fully resolved in Atwood's novels, the text still raises awareness for the pressing urgency of these issues in contemporary society.

Interestingly, animal consumption in Atwood's trilogy depict humans as inherently cannibalistic, while the non-human-ness of the Crakers is most tangible in their rejection of meat as food. While Crake's intentions of creating such a peaceful almost-human race may have been noble at heart, the novel makes very clear that the future of humanity lies not in a re-designing of genetic makeup, but in the negotiation of human normativities. Since the Crakers do not culturally distinguish between animals and humans in terms of edibility, they remain the genial scientist's curious experiment, which may exist alongside the human race but seems to be a poor substitute for humankind.

---

32   Atwood, p. 147.

Creation, Creator, and Cannibalism: the Craker Mythology

The biblical references to Snowman as prophet and Crake as creator lead to the necessary demand for a story of origin, a creation myth among the Crakers. Snowman takes on this role, not unwillingly, which highlights the inefficiency of the Crakers' imagination. They do not provide any kind of explanation for their existence themselves, but rely on the prophetic teachings of Snowman, which equips them with some understanding of their lives. While this clearly sets the Crakers apart from human communities, their interest in explaining their genesis suggests that the divide between Crakers and humans is growing smaller and smaller, without ever being fully closed. The Crakers are born, or better created, with an unchanging, instinctive sense of culture, while their ability to create, to imagine, and to evolve is strictly limited by design. Therefore, I would suggest that it is the inherent cannibalistic quality of humankind – which is reflected in anthropophagic rituals of the Eucharist in the Christian tradition – which allows for the unique and uncontested status of the human in Atwood's *MaddAddam* trilogy. Matters of diet and religion are thus intimately entangled and appear to provide the legitimation of human supremacy – despite the characters' attempts to challenge humanity with the development of new, intelligent, and potentially rival species.

Crake, who takes on god-like qualities in creating human life, explicitly rejects the idea of implementing religious thought in the Crakers' biological makeup: 'Crake was against the notion of God, or gods of any kind, and would surely be disgusted by the spectacle of his own gradual deification.'[33] However, as soon as the Crakers discover Snowman's explorations into the remains of the nearby city, they demand explanations of where and why he goes, and what he is doing there. Snowman cannot confront the Crakers with the true nature of their conception and simply explains:

> 'Where have you come from, oh Snowman?'
> 'I come from the place of Oryx and Crake,' he said. 'Crake sent me.' True, in a way. 'And Oryx.' He keeps his sentence structure simple, the message clear: he knows how to do this from watching Oryx […].
> 'Where has Oryx gone?'
> 'She had some things to do,' said Snowman. […][34]

---

33   Atwood, p. 104.
34   Atwood, p. 347.

The Crakers, having been part of a project conducted at the ironically named 'Paradice Complex' and, of course, unaware of their artificial creation, are now constructing a mythological narrative around their existence with the help of Snowman. The more Snowman is forced to explain to the Crakers where they come from, the more Crake is surrounded with an air of religious serenity: 'He [Snowman] is Crake's prophet now, whether he likes it or not; and the prophet of Oryx as well. That, or nothing.'[35] Crake as their maker acquires an almost deistic status, although his central aim was to edit out of their brains any need for religious ideals.

Just as much as Atwood's Crakers are in need of a creation myth as a place of spiritual and ideological origin, also the survivors of the 'Waterless Flood', the MaddAddamites, are creating their own mythology. Zeb, leader of the pre-apocalyptic activist movement 'Mad Adam' and now central figure in the post-apocalyptic survivor's enclave, goes from being the protagonist of eager children's stories (''He ate a bear once,' said Shackleton [...]. 'He ate the co-pilot. After he was dead, though.' Crozier said.'[36]) to be a central figure to the Crakers' mythology: 'We want to hear the story of Zeb. And the bear. The bear he ate.'[37] Estok aptly claims, 'Cannibalism evokes horror – and fascination'[38] thus conveying a feeling of staged monstrosity. Humanity's cannibalistic monstrosity is at the centre of Atwood's tale and is fuelled by the immanent incongruity of overpopulation and moral integrity. Indeed, the pre-apocalyptic society insists on clear boundaries between good and bad, although the factual difference between these categories has become more and more unstable. As a result, the boundaries of what is edible and what is inedible have been overthrown. The human boys confirm: '[Zeb] said bears look just like a man when they're skinned.'[39]

---

35   Atwood, p. 104.
36   Atwood, p. 105.
37   Atwood, p. 53.
38   Estok, p. 2.
39   Atwood, p. 53.

Conclusion: Cultural Cycles

Although Crake's aim of creating a pseudo-indigenous race was to surpass human mortality and prove that a peaceful human cohabitation is possible, the artificially created herbivoral instincts of the Crakers emphasize their non-human qualities. As the implied attribute of humankind is the grotesque ambiguity of being equally biologically edible and culturally inedible, the invalidity of such a distinction is all the more stressed with regard to the Crakers. The genetically mutated super-race unintentionally lacks the most basic human feature – that of edibility. They are therefore exempt from the causal ambiguity inherent in the human race, while at the same time also highlighting the humanity in the survivors all the more. Tina Pippin claims '[t]he path to paradise has dead bodies, evil monsters, heavenly messengers, [...], sacrifice, rape, cannibalism, plagues, and uncertainty'[40] while Magdalena Goldman, claims 'that Revelation is a misogynist, cannibal narrative [...]';[41] it seems logical that this very defining act of outwardly inhuman violence should be depicted during or after the destruction of known civilization, as it ultimately represents the repetition of a cycle of cultural development. In Atwood's post-apocalyptic world – which is as much the end of one era as the beginning of a new one – boundaries of human and animal, edible and inedible, and ultimately self and other appear to be no longer feasible. In a world of hybridity and newness the question of the origin and the necessity of a human nature remains questionable, as all that is created or has survived, all that seems artificial and all that is not, blends into a network of relationalities rather than into a rigid hierarchy of species.

Bibliography

Aigner, Andreas, Karl Pieper, and Herwig Grimm, "Post-Anthropocentrism' in Animal Philosophy and Ethics: The Disparity of the Prefix 'Post'", *HUMaNIMALIA,* 7 (2016), 56-83.

Atwood, Margaret, *Maddaddam* (London: Bloomsbury, 2013).

Atwood, Margaret, *Oryx and Crake* (London: Bloomsbury, 2003).

---

40 Tina Pippin, *Apocalyptic Bodies: The Biblical End of the World in Text and Image,* (London: Routledge, 1999), p. 83.

41 Marlene Goldman, *Rewriting Apocalypse in Canadian Fiction,* (Montreal; London: McGill-Queen's University Press, 2005), p. 85.

Atwood, Margaret, *The Year of the Flood* (London: Bloomsbury, 2009).

Avramescu, Cătălin, *An Intellectual History of Cannibalism* (Princeton: Princeton University Press, 2009).

Barker, Francis, Peter Hulme, and Margaret Iversen, *Cannibalism and the Colonial World* (Cambridge: Cambridge University Press, 1998).

Braidotti, Rosi, *The Posthuman* (Cambridge: Polity, 2013).

Calarco, Matthew, *Zoographies: The Question of the Animal from Heidegger to Derrida* (New York: Columbia University Press, 2008).

Derrida, Jacques, 'The Animal That Therefore I Am (More to Follow)', *Critical Inquiry,* 28 (2002), 369-418.

Derrida, Jacques, and Elisabeth Roudinesco, 'Violence against Animals', in *For What Tomorrow: A Dialogue* (Stanford: Stanford University Press, 2004), pp. 62-76.

Estok, Simon C., 'Cannibalism, Ecocriticism, and Portraying the Journey', *CLCWeb: Comparative Literature and Culture,* 14 (2012), 1-9.

Fausto, Carlos, 'Feasting on People: Eating Animals and Humans in Amazonia', *Current Anthropology,* 48 (2007), 497-530.

Foer, Jonathan Safran, *Eating Animals* (London: Hamish Hamilton, 2009).

Freud, Sigmund, *Die Traumdeutung* (Leipzig; Wien, 1900).

Goldman, Marlene, *Rewriting Apocalypse in Canadian Fiction* (Montreal; London: McGill-Queen's University Press, 2005).

Haraway, Donna J., *Simians, Cyborgs and Women: The Reinvention of Nature* (New York: Routledge, 1991).

Haraway, Donna J., *When Species Meet* (Minneapolis: University of Minnesota Press, 2008).

Hayles, Katherine, *How We Became Posthuman: Virtual Bodies in Cybernetics, Literature, and Informatics* (Chicago; London: University of Chicago Press, 1999).

Hulme, Peter, 'Introduction: The Cannibal Scene', in *Cannibalism and the Colonial World*, ed. by Francis Barker, Peter Hulme and Margaret Iversen (Cambridge: Cambridge University Press, 1998), pp. 1-38.

Kilgour, Maggie, *From Communion to Cannibalism: An Anatomy of Metaphors of Incorporation* (Princeton; Oxford: Princeton University Press, 1990).

Lawlor, Leonard, *This Is Not Sufficient: An Essay on Animality and Human Nature in Derrida* (New York; Chichester: Columbia University Press, 2007).

Miccoli, Anthony, *Posthuman Suffering and the Technological Embrace* (Lanham, MD: Lexington Books, 2010).

Pippin, Tina, *Apocalyptic Bodies: The Biblical End of the World in Text and Image* (London: Routledge, 1999).

Regan, Tom, *The Case for Animal Rights*. updated edn (Berkeley, London: University of California Press, 2004).

Ryder, Richard, 'Experiments on Animals', in *Animals, Men and Morals: An Enquiry into the Maltreatment of Non-Humans*, ed. by Stanley Godlovitch, Roslind Godlovitch and John Harris (London: Gollancz, 1971), pp. 41-82.

Singer, Peter, *Animal Liberation*. 3rd edn (New York: Ecco, 2002 [1975]).

Spiegel, Marjorie, *The Dreaded Comparison: Human and Animal Slavery.*

Rev. and expanded edn (New York: Mirror Books, 1996).

Steiner, Gary, *Anthropocentrism and Its Discontents: The Moral Status of Animals in the History of Western Philosophy* (Pittsburgh: University of Pittsburgh Press, 2005).

Villagra, Analaía, 'Cannibalism, Consumption, and Kinship in Animal Studies', in *Making Animal Meaning*, ed. by Linda Kalof and Georgina M. Montgomery (Michigan: Michigan State University Press, 2011), pp. 45-56.

Wolfe, Cary, *What Is Posthumanism?* (Minneapolis: University of Minnesota Press, 2010).

Elena Ogliari

HUMANS AS ZOO ANTELOPES AND WHITE SWANS: ANIMAL IMAGERY IN HENRY JAMES'S 'JULIA BRIDE'

Women as Commodified Animals

In the influential 'Why Look at Animals?' (1980), John Berger laments the disappearance of *real* animals in capitalistic societies insofar as animals have been subjected to a progressive cultural marginalisation: in our world, either they are 'co-opted' into the family as pets or they are turned into spectacles, as is the case with zoo animals. The process of marginalisation began in the nineteenth century in Western Europe and North America, whose faces were then being transformed by the combined shaping forces of urbanisation and industrialisation, to the point that, today, what inhabits a heavily anthropised world is the commodified animal.[1] Interestingly, Henry James offers a fictional representation of the irremediable loss of real animals in an urbanised environment in his 1908 tale 'Julia Bride'. The animals which populate the New York depicted in 'Julia Bride' are zoo animals, performing bears, or swans which beautify the Central Park Lake with their elegance. Since she has grown up in a capital-driven society, the tale's protagonist likewise cannot allude in her reflections but to animals culturally consumed in order to make a profit – like dancing bears, which constituted a major attraction at local fairs at the turn of the twentieth century. Yet, the logic of financial capitalism does afflict animals and women in equal measure in James's New York of the 1900s. Through a complex animal imagery, James comes to articulate the proximity between marginalised women and animals turned into spectacle, by juxtaposing the female body with the animal's and by drawing a parallel between the constraints imposed on performing animals – made unable to fulfil their potential capabilities – and the social norms to which women are subjected. Images of animals become metaphor of the protagonist's life and of women's lot at the beginning of the twentieth century.

---

1   John Berger, 'Why Look at Animals?', in *About Looking* (London: Writers and Readers, 1980), pp. 1-26.

'Julia Bride' was first published in *Harper's Magazine* between March
and April 1908 in two instalments, and it was later included by James into
the seventeenth volume of his own *New York Edition*, together with other
works of the late period such as *The Beast in the Jungle* (1903) and 'The
Jolly Corner' (1908). Like 'The Jolly Corner', it is set in contemporary
New York, but there is no ghostly element in this tale: it is the story of
the beautiful Julia, a girl whose chances to marry the rich and respectable
Basil French are jeopardised by her six previous engagements with just as
many gentlemen and her mother's two divorces. In order to emerge from
this impasse, Julia tries to persuade her mother's second husband and one
of her past fiancés to lie in her favour: whereas the former should take the
blame for the divorce, the latter should declare the utter uncompromising
nature of their relationship. Since her efforts are not to be crowned by
success, her very name comes to assume a mocking nuance: indeed, Julia
Bride will never be a bride.

'Julia Bride' is one of the tales James wrote after his prolonged
sojourn in the United States over the span of two years, 1904 and 1905.
He returned to his homeland after an absence of more than twenty
years: at the age of sixty-two, he found a country he did not recognise,
changed by mass immigration from Southern or Eastern Europe and,
above all, by industrialisation and the advent of capitalism. In a New
York best defined as in a state of frenzy, where old buildings were being
demolished to make room for futuristic skyscrapers, people seemed to
care only for money. This radically changed metropolis – so different
from the New York of his childhood he would have idealised later
as an Eden-like city in the autobiographical *A Small Boy and Others*
(1913)[2] – is represented in 'Julia Bride', where possession of goods and
financial security are the characters' main concerns.

In the microcosm of 'Julia Bride' everything is reduced to an economic
return. Not even the domestic dimension offers insurmountable barriers
against the harshness of the financial world, which seeps with its cold
logic into the institution of marriage. In the tale, marriage is a market
whereby young ladies offer themselves – their beauty, their persona –
in return for economic support. Materialistic motives are not alien to

---

2   James idealises the America of his childhood in the first volume of his
     tripartite biography. The metropolis is depicted as a 'cornucopia' for its
     inhabitants, who enjoy the products of 'the bucolic age of the American
     world'. Cf. Henry James, *A Small Boy and Others. A Critical Edition*, ed. by
     Peter Collister (Charlottesville: University of Virginia Press, 2011), p. 60.

Julia's choice to marry Basil, as the girl, who lives in a 'horrible flat [...] too near the East Side',[3] aims to marry outside of her class. In such a world, with marriage debased to a transaction between bride and groom, a girl strives to appear as the best choice to eager men and she is conscious that being good-looking means marketability. Yet, of crucial importance is also a woman's reputation: what she does or seems to do constitutes her history, a recurrent term in 'Julia Bride'.

Conscious that her outer appearance plays a vital role in defining how she should be treated and what should be thought of her, a woman is afflicted by an endless and partially self-imposed observation: besides being subjected to the oppressive gaze of the others, maybe of marriage rivals, a woman learns to be always aware of the image she projects since she watches herself being watched. Thus, a woman turns herself into an object, more precisely an 'object of vision',[4] and so does the protagonist, who is simultaneously transformed by the others' gaze into a permanent show, as if she were a caged animal in the zoo.

As Peter Buitenhuis first observed in 1970, 'Julia Bride' is punctuated with images related to the semantic fields of finance and zoo keeping,[5] the ones running parallel to the others and both concurring to form the complex imagery of the tale. Indeed, not only is 'Julia Bride' resonant with echoes of James's extended stay in the United States, but it is also a specimen of his late style, which is famously characterised by a consistent use of imagery.[6] In his late works, James thickens his prose

---

3   Henry James, 'Julia Bride', in *Complete Stories 1898-1910* (New York: Penguin, 1996), 661-696 (p. 683).

4   I draw this concept from John Berger's *Ways of Seeing* (1972), in which he analyses the tradition of the female nude in the European post-Renaissance painting to expose the reification of women in fine art, a reification which he notes also in present society insofar as women are reduced – and reduce themselves – to 'object of visions'. The notions of spectatorship and objectification return massively in 'Why Look at Animals', the essay collected in *About Looking*, which indeed may be considered as Berger's step forward in his study on the reifying gaze, here analysed in relation to zoo animals, subjected to an oppressive observation excluding reciprocity. Since Berger, not unlike James, identifies a similarity in the lot of zoo animal and women in *About Looking*, his insightful reflections concur to constitute the theoretical framework for the present article.

5   Peter Buitenhuis, *The Grasping Imagination: the American Writings of Henry James* (Toronto: University of Toronto Press, 2016), p. 231.

6   Cf. Robert L. Gale, *The Caught Image. Figurative Language in the Fiction of Henry James* (Chapel Hill: University of North Carolina, 1964) and R. W. Short, 'Henry James's World of Images', *PMLA*, 68, 5, 943-960.

with images of a very concrete nature with the intention to add extra meaning, and so the aim of the present article is to shed light on what the animal zoo imagery may reveal about the deeper meanings of the tale and about James's 'cosmology'. I draw the notion of cosmology from an article by Robert Short, who believed that an analysis of the recurring patterns of images can provide the lineaments of James's perception of the world – James's cosmology indeed – because by describing some aspects of experience with specific images the writer offers us insights into his own personal view of them.[7]

Admittedly, the employment of animal imagery is not peculiar to 'Julia Bride'. In 1953, by attempting a systematic classification of image patterns in James's later works, Short noted the frequent occurrence of terms belonging to the 'image-area' he denominated 'cage-beast' and pinpointed ornithology as a common source for drawing images.[8] With regard to ornithological references, he added that words such as 'bird', 'wing' and the less recurrent 'dove' and 'nightingale' are often used just out of practical need, insofar as they give prominence to structurally important passages, which might otherwise pass unnoticed. More consistent is the use of the cage-beast tropes, to which James often resorted to depict a world where maintaining or conquering individual freedom prove to be difficult tasks. The tropes of cages can be also used to portray a rigid society from which characters strive to escape.[9] In the novella 'In the Cage' (1898), which foregrounds the cage-beast imagery at the level of the title, the working-class heroine, an adolescent telegraphist, has limited opportunity either for upward mobility on the social ladder or to influence the course of her destiny. The cage thus comes to represent the many constraints that confine the telegraphist to a certain station in life.

Zoo animal imagery in 'Julia Bride' produces an analogous claustrophobic effect: the fences dividing animals from their spectators allude to social barriers or conventions that shut in (or shut out) individuals, preventing them from leaving their space of confinement. Yet, this is just one among the many effects James obtains through the layered image complex of 'Julia Bride'. The tale is dense with animal tropes: to use Berger's words, there are references to animals co-opted into the family, such as parakeets, as well as to those co-opted into

---

7   Short, p. 943.
8   Ibid., p. 943, 954.
9   Ibid., pp. 946-947.

spectacle, such as zoo or performing animals. To characterise Julia, James concocts a network of images peculiar to her only: allusions to animals exploited for entertainment or ornamental purposes are used to merely describe the protagonist, no other character is ever compared to wild animals in close confinement. More precisely, the narrator employs images of birds (parakeets or swans) with an ornamental function, while the zoo animal imagery pertains to Julia's meditations on herself.

Since James's characters, when talking about themselves through images, opportunistically draw them from the author's varied sources,[10] it is not surprising that Julia resorts to images of animals, yet opting for a subcategory of them: her choice to allude to zoo or ornamental animals to chart her own experience is coherent with James's depiction of a world shaped by greediness, which even permeates the treatment of animals. The peculiar animal imagery enables James to investigate the social and power dynamics regulating a woman's position and behaviour in contemporary society: Julia abides by the rules prescribing her exhibition to the others' gaze and the commodification of her beauty. Yet, brought up as 'the freeborn American girl' who may get 'engaged and disengaged not six times but sixty' (689) if she wishes, she finds herself at odds in a society where girls presenting themselves as merchandise do not arouse indignation, but where women's sexual drive remains a taboo. And in James's New York, the transgression of social mores, especially of those governing sexual matters, is harshly punished: since they pose a danger, transgressive women are marginalized for their difference, kept separated like predators from their preys at the zoo.

'A Parakeet of Precious Plumage': the Commodification of the Beautiful Animal

'Julia Bride' begins with an exchange of glances between the heroine and Basil French, the rich young man she aspires to marry. The incipit presents all the kernels on which James builds his narrative, as it introduces both the theme of spectatorship and the animal imagery by means of evocative allusions to birds. The passage is worth quoting almost in its full length:

---

10   Short, p. 946.

She had walked with her friend to the top of the wide steps of the Museum, those that descend from the galleries of painting, and then, after the young man had left her, smiling, looking back, waving all gaily and expressively his hat and stick, had watched him, smiling too, but with a different intensity – had kept him in sight till he passed out of the great door. She might have been waiting to see if he would turn there for a last demonstration; which was exactly what he did, renewing his cordial gesture and with his look of glad devotion, the radiance of his young face, reaching her across the great space, as she felt, in undiminished truth. Yes, so she could feel, and she remained a minute even after he was gone [...] She might now have been taking it afresh, by the testimony of her charming clouded eyes and of the rigour that had already replaced her beautiful play of expression. Her radiance, for the minute, had 'carried' as far as his, travelling on the light wings of her brilliant prettiness – he on his side not being facially handsome, but only sensitive clean and eager. Then with its extinction the sustaining wings dropped and hung. (661)

From the very beginning, James situates the narrative in the mind of his heroine, there captured in a moment of tense reflection. The narration is encumbered with a strong subjectivity: 'Julia Bride' is the story of the protagonist's growing awareness, her perception of the surrounding world as coloured by her feelings and filtered by her interpretations of past and present events. Endowed with glimpses into her mental processes, the reader soon becomes aware that Julia centres everything on the eye of the beholder: even when able to reciprocate the other's gaze, her emotions, thoughts, and actions are determined by what she sees refracted in the other's eyes.[11] She scrutinises Basil's expressions in order to have a confirmation of his devotion: as far as she spots it on her fiancé's face, she rises to an ecstatic state. She hovers in the sky as a bird whose precarious energy to fly ('sustaining wings') is rooted in the encounters of gazes. Reiterated few pages later, this image of elevation creates a striking contrast with the allusion to a swan in the tale's final, where the ornithological reference entails a shift from elevation to a descent into the abyss with the bird described floundering in engulfing waters.

---

11   In her 2001 monograph on the heroines of James's short stories, Donatella Izzo makes a similar point when she states that 'the gaze of the beholder is the permanent mirror where Julia can find herself by being reduced to an image'. Cf. Izzo, Donatella, *Portraying the Lady: Technologies of Gender in the Short Stories of Henry James* (Lincoln & London: University of Nebraska Press, 2001), p. 138.

Julia is presented as pretty aware of her appearance and image. This character exemplifies what Berger stated in *Ways of Seeing*, insofar as she is 'almost continually accompanied by [the] image of herself'; as the woman Berger described cannot help but 'envisaging herself walking or weeping' while she is walking across a room or mourning the death of her father,[12] Julia figures out the expressions that one after the other appear on her face. At the same time, the narrator drops hints about her consciousness of the effect her beautiful sight may produce on the beholder: an impression later corroborated by the explicit comment that she is 'lucidly conscious of the inimitable, the triumphant and attested projection, all around her, of her exquisite image' (668). She is so conscious that she has no qualms about speculating on her beauty to be successful: it gradually becomes clear that Julia perceives her beauty as a precious commodity on which to capitalise to appear more desirable to eligible men.

James presses the idea of beauty commodification into his reader's mind, by setting two-thirds of the tale in an art museum devoid of any aesthetic function and by resorting to zoo animal imagery. The author's juxtaposition of a painting museum with the zoo is not altogether surprising, if we consider the common features these two places share. In both cases, visitors' experiences rely on the sense of sight: there may be little difference between proceeding from cage to cage and stopping in front of a picture before moving to the next one.[13] In particular, Julia associates herself to either a painting hanging on the museum wall or to the zebras and antelopes in the city zoo. A condition of passivity is stressed in both cases, as Julia – envisioning herself as a painting – and zoo animals seldom reciprocate the other's gaze. Or, better said, their subjectivity is not deemed important and, therefore, neglected.

Notions of spectatorship, ownership, and profit characterise the deeper structures of Julia's world, to the point that the economic dimension is alien neither to the museum experience nor to the zoo institution. In the opening sequence of the tale, Julia is waving goodbye to Basil on the steps of the Metropolitan Museum. As Izzo observed, the museum institution is deprived of any aesthetic function: it is reduced to a mere building that figures as a meeting point for the characters, who give only quick looks at the pieces on display.[14] Julia walks into

---

12   *Ways of Seeing*, p. 40.
13   'Why Look at Animals?', p. 23.
14   Izzo, p. 139.

the painting wing to meet Mr Pitman, her mother's second divorced husband, and though he remembers her as an art lover, the paintings come to represent for her only 'all the things she [i]s vainly after' (675). The desire for possession prevails over seeking aesthetic pleasure *per se*, so that the paintings and statues embody the financial security she may attain only by marrying Basil. The artefacts at the Metropolitan Museum are despoiled by the characters of any sacredness – James's greedy New York society eschews the idea of sacredness associated with Art – thus becoming just objects to possess.

The paintings nonetheless figure as a touchstone of unanimously recognised beauty: Julia comes to the conclusion that she must look as beautiful as the paintings of the gallery to win Mr Pitman's feelings. Musing on her vital need to 'hang as straight as a picture on the wall' (666-667), she exposes her schemes for exploiting her radiant look as a picklock to penetrate her ex stepfather's heart: willing to provide herself with home and security, Julia knows that only through his intercession she might get herself purchased – like a picture – by her rich fiancé. Her lucid analysis of the situation also casts light on the reifying process the heroine forces on herself: she exhibits her beautiful appearance to the others' gaze, by reducing herself to an object of sight, whose distinctive quality is being confined to passivity ('hang', 'straight').

The impression of beauty degraded to a mere instrument is conveyed more powerfully when the protagonist compares herself to antelopes and zebras in a zoo. In James's lifetime, the establishment of zoos was justified by zoo owners' claims of presumed pedagogical functions: far from being a recreational activity, the visit to a zoo was said to promote knowledge and public enlightenment as Natural History museums.[15] If he ever believed in zoo owners' claims, James nonetheless offers a bleak depiction of these institutions in 'Julia Bride', by highlighting to what extent the beauty and rarity of animals are here viewed just as a warranty of future revenues. Significantly, he makes his protagonist envision herself as animals which had great exhibition value in his days, when antelopes and zebras constituted the essential basis for any general collection to attract zoo-going public.[16] As clear from the beginning, Julia and her mother boast an unordinary beauty, like 'remarkably

---

15   'Why Look at Animals?', p. 21. Cf. Harriet Ritvo, *The Animal Estate*, (London: Penguin, 1990), p. 5
16   Bob Mullan and Garry Marvin, *Zoo Culture*, (Urbana: University of Illinois Press, 1999), p. 70.

striped or spotted' animals: their taste in clothing ('so perfectly possessed of clothes') and radiant look ('so perfectly splendid') have been the 'points' on which they capitalised to elicit admiration, obtain favours and ensure themselves a successful social career (669). Profit is the ultimate goal of both a zoo entrepreneur and Julia: whereas the former strives to purchase exotic animals that spectators flock to see, the latter has been taught since early childhood that a pretty girl is endowed with 'inexpressible charms and quoteable 'charms'' (682) for her immediate employment in case of need. Julia's meditations on zebra and antelopes, interpolated with terms referring to stock-exchange transactions, are an example of how zoo animal imagery is tightly connected with money, to shed light on a world where relationships among people are perverted in the form of transactions and beauty, whether human or animal, is a capital[17] to use as stated by the dicta of financial capitalism.

Conceiving beauty as an instrument to ensure one's success is not a source of embarrassment to Julia. She has grown up in a world where a commodifying attitude to beauty is normally accepted. Julia's mother, Mrs Connery, capitalised on her daughter's appearance to gain favour with the judges when she got divorced from Mr Pitman. Mr Pitman, the man whom Julia is now asking for help, was the target of harsh criticism from the girl at the time of the trial for the divorce: when she was a child, Julia witnessed against her stepfather in Court, repeating insubstantial accusations she had been taught by her mother. Due to her mother's interference, Julia does not find it inconvenient to approach him after so many years; indeed,

> *She* at least had never divorced him, and her horrid little filial evidence in Court had been but the chatter of a parakeet, of precious plumage and creak, repeating words earnestly taught her and that she could scarce even pronounce. (666)

The reader is again presented with the protagonist's thoughts, here depicted in an attempt to justify her asking for Mr Pitman's help. The above quotation also attests to Julia's peculiarity to talk about herself by resorting to images of animals: in the given case, she compares herself with a parakeet, a tropical pet which enjoyed enormous popularity in middle-class households at the turn of the century. The comparison is a quite obvious one, since parrots and the like usually represent people who

---

17   Izzo, p. 142.

repeat someone else's words without thinking about what they stand for, but the reference to the bird's beautiful plumage and creak adds further meaning, drawing attention to the utilisation of beauty for practical purposes. The child's loveliness was supposed to exert a favourable impression on the members of the jury to the detriment of Mr Pitman.

The allusion to the parakeet blatantly evokes negative associations in the reader's mind, because Julia's identification with the bird is fraught with painful memories of exploitation. The employment of animal imagery marks momentous events in Julia's life, which cast light on her subjection – self-imposed or not – to the rules of a society in which she does not perfectly fit. All the animal tropes punctuate the protagonist's descent into the abyss of complete submission to a set of norms that will impede her marriage. A negative note colours also the image of elevation in the opening sequence, because it represents an act of self-reification, but it is less striking: when described ascending to the sky under the adoring eyes of Basil, Julia seems to be bound to victory. In fact, James toys with his readers' expectations, cultivating the illusion of Julia's marriage to Basil and, building his narrative according to principles of symmetry, he counterbalances the images of flight with images of birds sinking in water. Ironically, the end begins when not only is her beauty at its peak, but she also conquers the genuine admiration of those persuaded of her moral superiority.

Since childhood, Julia has been taught to rely on her beauty to succeed in life, and her good look indeed has given her many satisfactions. Not only had she many suitors, but she is flattered by people's stunned reaction to her beautiful appearance, especially women's: unfailingly aware of her own image, Julia indulges in glimpsing admiration in other people's expressions as they are confronted with her sight. With Mr Pitman this inclination turns out to be the cause of her failure. Mr Pitman refuses to lie in Julia's favour, because he wishes to marry the rich widow Mrs Drack, and the only obstacle to their union consists of Mr Pitman's former marriage, as the woman disapproves of divorce. The only reasonable way out is to ask Julia the exact opposite of what the girl wanted from him, i.e. to declare that the divorce was only Mrs Connery's fault.

Although she is unwilling to aid him, Julia ends up complying with her interlocutor's requests because she falls victim to her awareness of the image she projects. When she meets Mr Pitman's fiancée, she cannot but enjoy her reflection in the widow's eyes, as she recognises that her own is 'unmistakeably the most dazzling image Mrs Drack had ever beheld' (680). Desiring to impress Mrs Drack even more, Julia takes a course of action which is the most unexpected in her society: she sacrifices her own interests – she jettisons the selfish logic that regulates her world – to tell the truth for Mr Pitman. Her action increases her beauty in Mrs Drack's eyes: Julia may glimpse admiration in the widow's glance, intensified by the latter's conviction that Julia's outer beauty is just the reflection of the inner one. This recognition, the consequence of an exchange of glances, tickles Julia's narcissism, this latter thus indulging so much in the other's gaze that she reaches a state of ecstasy, not dissimilar to the one experienced with Basil on the steps of the Metropolitan, albeit stronger. Lingering in the effect produced by her sight, Julia spreads her own wings ('swan-like'[18]) to rise and soar as a bird into an elevated state:

> She measured every beat of her wing, she knew how high she was going and paused only when it was quite vertiginous. Here she hung a moment as in the glare of the upper blue; which was but the glare– what else could it be?– of the vast and magnificent attention of both her auditors, hushed, on their side, in the splendour she emitted. (681)

It is a moment of intense emotion which the reader witnesses through the filter of Julia's eyes: the rhetorical question concurs to place the narrative in the consciousness of the protagonist, who relishes observing the couple as they stand still to stare at her with rapt admiration. Despite the almost overwhelming emotion she feels, in her ascent towards the blue above, Julia manages to exert a certain degree of control over the process by pausing when the height becomes vertiginous. The second part of the tale ends with this image of the heroine's apparent triumph; but, though the abilities of observation and reflection are equally well developed in the tale heroine, this time she fails to recognise that her stopping before dizziness only gives her the illusion of holding the reins of the situation. She does not feel impotent yet, but it is indeed impotence that will soon seep into her mind. The images of ecstatic hovering are to be outnumbered by images of abysses and confined spaces, where the animals are unable to free themselves from the constraints others have imposed onto them.

---

18 'Julia Bride', p. 682.

James's Zoo Animal Imagery: an Insight into his own Cosmology

Having lost the support of Mr Pitman, Julia turns for help to Murray Brush, a former fiancé just returned from Europe. She hopes the Old World has refined his manners and transformed him into a gentleman: if it is so, Murray will help her by telling Basil that their engagement did not have any objectionable consequences. Basil has stepped into Julia's world with different notions of right and wrong, spurring the girl to conceive her six engagements in terms of impropriety (668). In the very surname of Julia's seventh suitor, French, there are embedded allusions to conservative Europe with its codes of honour and chastity.[19] For Basil, who is 'awfully like a high-chaste Englishman', and his family marriage is 'a great matter'. The 'serried Frenches' (668-669) have sexually repressive views about marital issues: as defenders of traditional marriage they uphold old values of female domesticity and sexual purity, which are threatened by the restlessness of women such as Mrs Connery and her daughter.

A range of completely different conventions prevails in Basil's circle even in relation to beauty. In the past, Julia saw her beauty being utilised as an instrument by herself and by the others. The 'exhibit' of the parakeet at the trial was not an isolated fact, because 'there was no one, all the while, who wasn't eager to egg you on, eager to make you pay to the last cent the price of your beauty' (667). Yet, whereas her past successes were due to her physical appearance, a pretty face is not enough to convince Basil to marry her: on the contrary, she would rather lose 'a pound or two of [her] treasure' in change of a 'less published personal history' (668). History stands for reputation, which in Basil's market marriage has a greater exchange value than beauty.

Her acquaintance with Basil's world provides Julia with a new perspective from which she can judge the society to which she belongs, coming to realise that her history has inevitably shaped the relationship between her and the others. Mrs Connery and Julia's beauty has always attracted other people's gaze, but it is their history, made of two (almost three) nullified marriages and six engagements, the key factor in their transformation into spectacle. The two women are reduced to curiosities, an unenviable status enticing voyeuristic watching from spectators who are eager for cheap thrills. As the story progresses, Julia increasingly sees that their acquaintances are prone to

---

19   Buitenhuis, p. 227.

admiring them, talking to them, through the rails, in mere terms of chaff, terms of chucked cakes and apples – as if they had been antelopes or zebras, or even some superior sort of performing, of dancing, bear. (668)

In the tale, the activity of watching is never harmless: on the contrary, James further despoils the act of observation of any pretence of innocence as he gives prominence to the exploitative associations, coloured with constraint and oppression, which his zoo animal imagery may engender. Mother and daughter are subjected to a constant scrutinising gaze. These human antelopes and zebras are conveniently caged to be exhibited and kept under surveillance by the same spectators who assume the role of guards by 'fencing them [...], and so not only shutting them out from others, but mounting guard at the fence, walking round and round outside it to see they didn't escape' (668). Like zoo animals and performing bears, Julia cannot escape the indignity of being exhibited against her will.

In her examination of imagery in James's late works, Susan Commanday juxtaposes the above passage on zebras and antelopes with an extract from 'The Two Faces' (1900), in which Lady Gwyther is said to be 'overloaded like a monkey in a show' to demonstrate that James nourished empathy for animals forced to exhibit themselves and perform before a crowd.[20] Admittedly, when applied to the case of Julia, the whole gamut of animal imagery is not only bestowed by the author with more intense evocative power, but it also seems to be coloured by a suffused feeling of empathy for the animals reduced to objects of sight. Yet, a perusal of his correspondence and private documents does not suggest that James believed that animals have rights. Neither did he express his mind on those public animal welfare issues that sparked heated debates in England between the late nineteenth and early twentieth century, which questioned the morals of vivisection and shows featuring animals.[21] On second thought, the reader realises that James harboured sympathetic feelings for those individuals who face the same unwanted exposure as zoo or performing animals. At the most, if he ever sympathised with zoo or performing animals, it was because he projected his fears onto them. Many years before writing

---

20   Susan N. Commanday, *Imagery in Henry James's Late Stories (1898-1910)*, (New York University: Ph.D. Dissertation, 1973), p. 232.

21   Cf. David A. H. Wilson, 'Politics, Press and the Performing Animals Controversy in Early Twentieth-Century Britain', *Anthrozoös*, 21, 4, 317-337

'Julia Bride', he had found himself exhibited before a gaping public at the premiere of his 1895 play *Guy Domville*, which ended with the helpless James being booed from the gallery as he bowed onstage. There followed a period of dejection for the writer, who saw his dream of becoming a professional playwright being shattered to pieces. In a letter of January 9 to his brother William, he recalls the nightmarish experience charging it with Dantesque connotations, as he remembers when his friends assembled in the theatre

> waged a battle of the most gallant, prolonged and sustained applause with the hoots and jeers and catcalls of the roughs, whose roars (like those of a cage of beasts at some infernal 'zoo') were only exacerbated (as it were) by the conflict.[22]

James is not adamant about the separation between animal and humans, because even the latter may be treated as entertaining living exhibits: Julia and her mother resemble animals in cages as all their acts are being examined and commented upon. When James visited the London Zoo in 1869[23] and the Central Park Zoo during his stay in the United States, both places architecturally resembled old menageries with the animals exhibited behind bars in cube-like cages, which ensured their perpetual visibility: the notion that the animal should be entitled to be off-exhibit, as it wishes, was completely neglected.[24] Like those animals, Julia cannot absent herself from the others' gaze: the neighbour Mrs George Maule, for instance, scrutinises every movement of Julia while keeping a detailed record of her love affairs to discredit her, hoping that Basil, disgusted by the heroine's frivolity, will marry one of her four daughters (675).

Yet, the reader may perceive a more disturbing tinge in the protagonist's identification with antelopes and zebras in the zoo, because her parallelism brings to the surface fears of containment, control and separateness. The author variates on the theme of zoo keeping by expounding the dynamics of power embedded in the process of separating and confining animals. At the turn of the twentieth century, animal inferiority justified their cultural consumption – the

---

22   Leon Edel, *Henry James: The Treacherous Years, 1895-1901* (Philadelphia: Lippincott, 1969), p. 79.
23   Cf. Henry James, *Letters 1843-1875*, ed. by Leon Edel (Cambridge: Harvard University Press, 1975), p. 93.
24   Mullan, p. 70.

instrumentalisation in anthropocentric institutions such as zoos or the exploitation in circus and fairs – and their separation from humans: not being considered fellow creatures, their lack of freedom was no matter of concern for zoo entrepreneurs. The zoos epitomised the traditional Western thinking about animals, which purported human beings at the top of the hierarchy. But by dividing the preys from the predators, the mammals from the birds, the traditional zoo embodied not only the hierarchy between animal and human beings: its obsession with classification and rigid barriers between species easily made it an analogous of social divisions in human societies. In a fictional depiction, the human-animal hierarchy of the zoo may symbolise the hierarchy among human beings.

'Julia Bride' deals with issues about human society worrying James, who, in this tale, represents the socially allegorical resonances of zoos. Julia's acquaintances recast, at an imaginative level, the physical barriers and structures of separation that create an unbridgeable divide between humans and animals at the zoo. The spectators take on the role of guards who erect barriers that prevent Julia from moving freely in society: her reflection on the others' mounting fences around her has clear claustrophobic overtones. Moreover, as zoo animals are forced to the unnatural role of living exhibits, which prevent them from fulfilling their potential capabilities, Julia's peers confined her to the role of the beautiful object to be exhibited, whose power of attraction is increased by her six engagements. The heroine cannot step out from the role others have created for her, but she is bound to be 'nothing *but* pretty' (667), while prompting curiosity for her record of suitors. More importantly, this frame in which she is caged annihilates Julia's identity by reducing it only to a difference existing between her and the others – a difference which shuts her out from many sectors of society. Julia is being marginalised by both the Frenches and her world, because of the past six engagements.

From the point of view of her background social group, Julia is the specimen of a different species because her engagements were not motivated by materialistic reasons: her fiancés were not rich, so hers was a passion triggered by pure sexual pleasure. She and her mother 'amused themselves' (666) by collecting suitors. Yet, a concession to pure pleasure is inadmissible in a world where marriage is the vehicle to reach financial security or status: a woman's sexual desire should always be in the sanctioned form of a quest for wealth.[25] Her transgression of

---

25   Izzo, p. 143.

the norms regulating the sphere of sexuality makes her a potential threat
to be caged. By confining her to the fixed role of a curiosity, of the
pretty but eccentric girl, they have the impression to contain the danger
she posits with her disregard for rules.

Instead, the Frenches see in the succession of suitors a consequence
of Julia's moral slackness, so that Julia belongs to a different species
to isolate. Upholding aristocratic definitions of the woman's role in the
married couple, the Frenches strive to protect themselves against new
permissive standards about sexuality – personified by 'the freeborn
American girl' (682) – which they deemed as the symptom of moral
decline. Extremely sensitive to gossip, they would not hesitate to erect
barriers against Julia if her story were confirmed. The Frenches are
presented as a closed group, as signalled by the claustrophobic echoes
resonating in the every allusion to them: not only are they 'serried' as a
phalanx, but Basil's relatives also appear as barricaded 'in their fortress',
which is seldom 'accessible' to those who do not conform to their codes
of honours and respectability (669; 695). Julia is confronted with the
'gate [...] of the grand square forecourt of the palace of wedlock'
insistently kept closed by the Frenches. The heroine has always been
denied access to the domain of Basil – the gate significantly has 'such
stiff silver hinges' complicating her 'introduction to his precinct' (669)
– and her admittance will not be granted, unless she proves her chastity.

Julia is as much subjected to the Frenches' codes as to the norms
regulating her background society. As Izzo noted, the heroine's
'subjection to the morals of others is complete': her precarious financial
situation prevents her from contrasting those moral conventions that
obtain in Basil's world, because cleaving to them is the *conditio sine qua
non* to marry Basil and attain wealth. Moreover, unable to comprehend
the 'real terms of her subjection', Julia pinpoints the causes of her current
marginalisation in her past mistakes and on contingent circumstances[26].
She especially blames her haphazard upbringing: as dictated by her
business-oriented society, she has learnt to instrumentalise her beauty
and exploits the others, but she failed to exert control over her reckless
social career ('the disgusting humiliating thing'[27]), being unaware of the
crucial importance of reputation.

At last, Julia recognises the impossibility to reverse the process
of marginalisation while talking with Murray Brush at Central Park,

---

26   Izzo, p. 145
27   'Julia Bride', p. 688.

where she reaches the baleful climax of her growing awareness. Central Park provides the setting for the tale's conclusion and acts as a locus of anxieties, where the protagonist plays out her own fears of surveillance and her conflicts with the past. Images of confinement recur obsessively in the conclusion, as the protagonist grows aware that she will be unable to escape from the trap she has fallen into. At the park, she cultivates the illusion of absenting herself from the others' gaze by choosing to meet Murray 'in a sequestered alley of the Park': ironically, she opts for secluded paths where she used to meet Murray at the time of her engagements. The two 'tr[ead] then afresh their ancient paths' 'haunted' by the 'general echo of her untrammelled past' (683).

As her choice points out, Julia cannot free herself from her scandalous past, which forms an unbreakable cage around her. The girl perceives that her efforts will be vain as soon as she starts doubting the genuineness of Murray's offer to help her: she senses that her former fiancé, an ambitious social climber, will aid her only so far as his help will facilitate his entrance into the domain of the Frenches. Julia cannot but recognise her defeat, 'her now certain ruin' (696), which she envisions as sinking in water. The mental process leading the heroine to the final revelation is marked by a shift from the images of flight to those of submersion. She remembers when 'two days before' she 'floated' in the sky 'swan-like'. But now, before her, Julia sees a 'rising tide' whose force she cannot resist: since 'cold swish of waters [are] already up to her waist' and will 'soon be up to her chin' (691), the swan is engulfed into the abyss. She did not manage to contrast the dynamics that relegated her to the role of the outcast or of an object of curiosity, and it is impossible to her to delete her own history: thus, she can only face a destiny whereby she will always be the eccentric unmarried girl who had seven engagements. With a typical swift act of reflection, the heroine realises the futility of denying the defeat as she knows 'inevitable submission, not to say submersion, as she ha[s] never known it in her life' (691): with the close juxtaposition of 'submission' with 'submersion', the author conveys in few words the thematic kernel of his narrative – the struggle against rigid social mores, which he depicted through the employment of animal imagery. And in the face of the last revelation about her defeat, Julia surrenders to the tide submerging her.

Bibliography

Berger, John, *Ways of Seeing* (London: Penguin Books, 1972)

Berger, John, *About Looking* (London: Writers and Readers, 1980)

Buitenhuis, Peter, *The Grasping Imagination: the American Writings of Henry James* (Toronto: University of Toronto Press, 2016)

Commanday, Susan N., *Imagery in Henry James's Late Stories (1898-1910)*, (New York University: Ph.D. Dissertation, 1973)

Edel, Leon, *Henry James: The Treacherous Years, 189 -1901* (Philadelphia: Lippincott, 1969)

Izzo, Donatella, *Portraying the Lady: Technologies of Gender in the Short Stories of Henry James* (Lincoln & London: University of Nebraska Press, 2001)

James, Henry, *Letters 1843-1875*, ed. by Leon Edel (Cambridge: Harvard University Press, 1975)

James, Henry, 'Julia Bride', in *Complete Stories 1898-1910* (New York: Penguin, 1996), 661-696

James, Henry, *A Small Boy and Others. A Critical Edition*, ed. by Peter Collister (Charlottesville: University of Virginia Press, 2011)

Mullan, Bob and Garry Marvin, *Zoo Culture*, (Urbana: University of Illinois Press, 1999)

Ritvo, Harriet, *The Animal Estate*, (London: Penguin, 1990)

Short, R. W., 'Henry James's World of Images', *PMLA*, 68, 5, 943-960

Wilson, David A. H., 'Politics, Press and the Performing Animals Controversy in Early Twentieth-Century Britain', *Anthrozoös*, 21, 4, (2008), 317-337

Daniela Maria Hirsch & David Gaehtgens
DISPLAY
AN EXPEDITION INTO GAZING

Expedition Report

An artistic field study December 2014 – January 2016
Destination: Zoological Garden Berlin, Germany
Concept, implementation and evaluation: Daniela Maria Hirsch &
   David Gaehtgens
Permanent documentation: www.gaehtgenshirsch.net/surrogates
Status of expedition: Returned
Status of report: May 2017, preliminary

Abstract

Over the course of one year we have gathered longterm video recordings of roaming humans gazing at displayed animals in the Berlin Zoo, which is the oldest German institution of its kind. The setting of the zoo focuses prototypically on central aspects of the human posture towards non-human animals, but also on human organization itself: the distribution of participation, the access to and development of capabilities and the effects of shared narratives and agreements.

The zoo's phenomenology carries both historical and current information on human positions. Here, the viewing direction towards co-animals is visibly manifesting in space and embedded in the overall concept of management and application for use. Therefore the tool of the camera is directed at us humans, the other animals are mostly absent from the recordings.

The study is based on a series of full-day field trips into a semi-public institution which features models of habitats for 18,662 animals from 1,380 species on a total area of 81.5 acres (as at 2015).[1] It generated

---

1   https://www.statistik-berlin-brandenburg.de/produkte/Jahrbuch/jb2016/
excel/JB_201604_BE.xlsx, accessed on 29 May 2017

a visual compendium which consists of three chapters: 'Looking In', 'Looking Out' and 'Safari'. It aims to explore and evaluate the process of gazing as a significant marker of human interests and intentions.

Expedition Review

Background

The human species has added its own peculiarities to the animal world. One unique feature is the broad array of customs mankind holds central to social organization. Human animals have developed by living as associates. The concept of collaboration is and has been central to the maintenance and evolution of the human concept. Connected to this is a manifold spectrum of social arrangements and commitments.[2] The packaging and coding of this information varies greatly among human groupings.

Customs are complex inventions which serve various purposes. One of them is the dealing with otherness.[3] The correlation between likeness and unlikeness, between similar and dissimilar is attributed with great dynamics. It concerns strategies for the provision of survival, but moreover humankind is interested in abstraction, elementary meaning and explanatory systems.[4] These renderings have an intrinsic impact on decision making and actions. Therefore we consider habitual practice as a central force within human aspiration. To examine this field, we have chosen a starting point in our own cultural breeding.

Research Area

Our biographical backgrounds lie in central Europe. Here, as in other parts of the global human settlement, dealing with distinction is

---

2   Desmond Morris, *The Naked Ape* (Frogmore: Triad/Mayflower Books, 1977), p. 21: 'The forest ape that became a ground ape that became a hunting ape that became a territorial ape has become a cultural ape' […].
3   John Berger, *Why Look at Animals?* (London: Penguin Books, 2009), p. 16-17: 'This – maybe the first existential dualism – was reflected in the treatment of animals. […] The parallelism of their similar/dissimilar lives allowed animals to provoke some of the first questions and offer answers'.
4   e.g.: explanation of man's origin can be found in human cultures around the world

embedded in the effort to acquire and maintain meaningful perceptions in an apparently chaotic world. But here the interest and curiousness in other beings of nature was influenced by a specific change in the world view: gradually Europeans stepped out of the cosmological experience in which everything was part of a grand divine plan and into a more and more self-empowered position.[5] Human curiosity drove towards an emancipated quest for detail and context. If it was not a divine system of logic, then which correlations could there be? And if no god was ruling the course of events, why not strive for an autonomous human regulation?[6]

This change in self-esteem is reflected in the intercourse with objects of the natural world. The mapping of the world though travelroutes and their abstract representations via graphical systems did not only give access to many findings of otherness. This otherness stirred the imagination but also challenged the intellectual claim to find a grid through which occurrences could be handled. The system of mapping includes the method of collection. But not only was immaterial information gathered and stored, actual natural objects of all kinds were also transported and rallied. Initially, those findings represented both the powerful quest in realms of the world and the strange entertainment the universe holds.[7] The early storage rooms for these collected objects in the 14th century were called *Wunderkammer* (cabinets of curiosities) and they represented the fascinating experience of the world. Parallel and even preceding to this is the *ménagerie* – 'une établissement de luxe et de curiosité': a part of a garden or a park in which a collection of captive animals, frequently exotic, are kept for display.[8]

In the further course however, the indoor display of collected items expanded from a prestigious demonstration of monetary and imperial power to the question of encyclopedia: how can the collection of

---

5   Paul Virilio, *Der negative Horizont*, trans. by Brigitte Weidmann (Munich: Carl Hanser Verlag, 1989), p. 38: 'Wie hat man im Tier das Vehikel, in seinem Fleisch den Motor vermuten können?'

6   e.g. Galileo Galilei's life and work hold many examples and illustrations.

7   Eric Baratay, Elisabeth Hardouin-Fugier, *Zoo – Von der Menagerie zum Tierpark*, trans. by Matthias Wolf (Berlin: Verlag Klaus Wagenbach, 2000) p. 29 (citation from Krzysztof Pomian, *Der Ursprung des Museums. Vom Sammeln* (Berlin, Wagenbach, 1998), p. 13-70): 'Diese Kuriositätenkabinette, die nur selten spezialisiert waren, stellten sowohl vom Menschen geschaffene *artificialia* zur Schau […] als auch natürliche Objekte, sogenannte *naturalia* […] als Kondensierungen der wahrnehmbaren, erkennbaren Welt'.

8   Ibid p. 41: '1662-1664 entstand dann in Versailles eine Ménagerie, die ausschließlich für exotische, seltene und ungewöhnliche Tiere bestimmt war'.

objects from the natural world become part of a reference system, generating and endorsing explanations? The wonder room turned into a *Kunstkammer* in which the magic of nature should be acquired for specific applications and human intention.[9] Work on explanation was always escorted by representation, the room itself served as a three-dimensional translation of achievement. The objects on display are no longer mere witnesses but have become accomplices in conceptual modelling. Seeing and thinking are manifested together. Ultimately this drive leads to the museum as a body of competence, a place to represent and connect seemingly valid models of realities.[10]

The concept of offering a view through organized presentation implies the distribution of positions.[11] With the development of the modern museum there comes the position of a manifestly objective institution to represent facts in categories. These categories aim to establish rational knowledge in the context of larger collections. As the

---

9   Horst Bredekamp, *Antikensehnsucht und Maschinenglauben* (Berlin: Verlag Klaus Wagenbach, 2000), p. 53, after Francis A. Yates, *Aufklärung im Zeichen des Rosenkreuzes* (Stuttgart: Klett, 1975), p. 203-208; R. F. Ovenell, *The Ashmolean Museum 1683-1894* (Oxford: Calendon Press, 1986), p. 1-30; Samuel Quiccheberg, *Inscriptiones vel tituli theatri amplimissimi, complectentis rerum uniersitatis singulas materias et imagines eximias [...]* (München, 1565) 'Palissys und Andreaes Vorstellungen, die Kunstkammern nicht nur als passive Sammlungen, sondern auch als aktives Labor zu nutzen, entsprach der prometheischen Praxis, Sammeln, Forschen und Gestalten als Einheit zu begreifen. Schon Quicchebergs 'Inscriptiones' lassen die Einteilung der Kunstkammer in die Beschreibung der zugehörigen Werkstätten münden. Für die Fürsten und Kaiser bedeutet eine aktive Beteiligung an der Forschung [...] einen besonders sinnfälligen Ausweis ihrer absoluten Herrschaft, die nicht allein ihre repräsentative Würde, sondern und vor allem eine aktive Kontrolle der Außenwelt betonte'.

10   Bredekamp, p. 32-33: 'Als exaktes, ordnendes Instrument betont die Uhr den imperialen Gestus [...]. Die Geschicke der Welt wie nach einem verbindlichen Zeittakt zentral bestimmt laufen zu lassen, war, [...] zum sinnfälligsten Ausweis reibungsfrei gelenkter, absolutistischer Herrschaft geworden [...].' p. 34-35: 'In der dritten Abteilung sind die drei Naturreiche des Animalischen, Vegetabilen und Mineralogischen systematisiert: [...]. [...], wobei das Schwergewicht auf die Ordnung und nicht auf die Verarbeitung der Objekte gelegt ist'.

11   John Rachman, *Foucaults Kunst des Sehens* in *Imagineering*, ed. by Tom Holert (Cologne: Oktagon Verlag, 2000), p. 47: 'Es ist diese begriffliche Neuorganisation oder 'Verräumlichung', wodurch die Naturgeschichte 'nichts anderes als die Bennennung des Sichtbaren' wurde. [...] Das räumliche Schema einer Wissensform ist nicht nur von den darin vorkommenden Theorien verschieden; oft geht es ihnen voraus und ermöglicht sie erst'.

research focuses on the broad collection of examples and occurrences, so should the visitors follow this activity – such as James Smithson longed for the founding of 'an establishment for the increase and diffusion of knowledge among men.'[12] With the Smithsonian Institute there is concept proof of this idea. But what are the means on the viewer's side?

The act of looking at a display is intended to be of a communicative nature, yet the procedure of gazing carries special attributes. We aim to look at these in regard of knowledge production.

Expedition Area

Long before the collecting of other animals, there is the keeping of them for nutrition or muscle power. This marks a decisive point in human development. The taming and breeding of certain species for specific purposes started around 12,000 years ago.[13] Not only is this a boost in self-empowerment, additionally there is a prothetic effect mankind achieved through access to more-than-man-power. This is prolonged in the association with certain types of animals to gain prestige or show prestige via trophies. The collection of live animals in aristocratic enclosures also could – like the *Wunderkammer* – encircle a little world as a seeming representation of the whole world.[14] And like in the 'real' world it was also about hunting them, sometimes in sophisticated settings of landscape architecture which equally enabled viewing and observation. There always was a special focus on exotic animals – foreign, from the outside and distinct.[15]

As the appropriation of the larger natural world progressed, this reflected on human exhibition culture: Zoological societies and gardens have emerged as a product of scientific progress and

---

12   https://siarchives.si.edu/history/james-smithson, accessed on 29 May 2017
13   Greger Larson and Dorian Q. Fuller, 'The Evolution of Animal Domestication', *Annual Review of Ecology, Evolution, and Systematics,* Vol. 45, 2014, pp. 115-136
14   Baratay, p. 44: 'Mit Hilfe des Gartens ließ sich der [...] Wunsch nach einer Rückkehr zur kosmischen Fülle [...] verwirklichen, [...]. Gleichzeitig sollte er jedoch auch hervorheben, mit welchen Elementen und durch welche Maßnahmen dieser Mikrokosmos – das Abbild des idealen Kosmos – geschaffen worden war [...]'.
15   Ibid, p. 37: 'Diese Tiere aus dem Ausland – Rassen die wenig bekannt und insofern ungewöhnlich waren – boten den Adligen eine willkommene Gelegenheit, sich als Landwirte zu fühlen, ohne deshalb vulgär zu erscheinen'.

application.[16] They carry within themselves core ideas of modern western civilization. Also they reflect on the position of industrialized man in and towards the natural world.[17]

This is why we chose a zoological museum as destination for our expedition. Because we wanted to start studying from our own immediate habitat, we decided on the zoo in our resident city of Berlin. It happens to be the oldest zoo in Germany. It was founded after the physicist and explorer Martin Hinrich Lichtenstein persuaded the prussian king Friedrich Wilhelm 4th in 1841 to not only give him a loan, but also to dedicate a part of his private pheasantry free of cost to this mission. Alexander von Humboldt, the prominent naturalist explorer, was a strong supporter of this cause and influenced the king, enabling the opening to a selected public in 1844. The older and ambitious predecessor in London was the first ever to use the term Zoological Garden when it opened for scientific study in 1828. Today the Berlin institution is the zoo with the highest number of domiciled species worldwide.

Since the project of becoming a civilizee is one of the human animal, we decided from the beginning to look at our own species.[18]

---

16   Ibid, p. 127 after Georgette Légée, *Le Museum sous la Revolution, l'Empire, et la Restauration* in *Histoire de l'enseignement de 1610 à nos jours* (Paris: CTHS, 1974), p. 752: 'Mit der Eröffnung der Menagerie im Jardin des Plantes waren die Utopien Bacons und Leibniz sowie die Vorstellungen der Gelehrten des 18. Jahrhunderts Wirklichkeit geworden. Es handelte sich um die erste Institution ihrer Art, die von Wissenschaftlern und ausschließlich für sie eingerichtet worden war. [...] Mit ihren ehrgeizigen Programmen im Bereich der Anatomie, Physiologie, Klassifizierung, Verhaltensforschung und Akklimatisation [...] leisteten sie wahre Pionierarbeit. [...] In Berlin [...] beispielsweise arbeitet man immer wieder mit Universitätsinstituten zusammen. Wissenschaftliche Forschung galt jetzt überall [...] als das Gebot der Stunde [...] das da lautete: Inventarisierung, Akklimatisierung und Domestizierung zum Zwecke der Ausbeutung'.

17   John Berger, *Why Look at Animals?* (London: Penguin Books, 2009), p. 30: 'Public zoos came into existence at the beginning of the period which was to see the disappearance of animals from daily life. The zoo to which people go to meet animals, to observe them, to see them, is, in fact, a monument to the impossibility of such encounters. Modern zoos are an epitaph to a relationship which was as old as man. They are not seen as such because the wrong questions have been adressed to zoos'.

18   Desmond Morris, *The Naked Ape* (Frogmore: Triad/Mayflower Books, 1977), p. 26: 'The sensory equipment of the higher primates is much more dominated by the sense of vision than the sense of smell. In their tree-climbig world,

Furthermore we expected to look at groups and community dynamics in a semi-public space. Our visits were full-day experiences.

The area itself was known to us from previous visits – some private ones many years ago and two recent ones for researching purposes in connection with the expedition. In 2015 the Berlin Zoo had 3.3 million visitors, making it the most visited zoo in Europe.[19]

Expedition Dates

On 29 December 2014 we bought an annual ticket to the Berlin Zoo for our first expeditional visit. Because it was the holiday season, a friendly staff member made our passes start on 1 January 2015, giving us an extra of 2 days. We made 29-day trips there over the whole year in all seasons and weather conditions. Our last visit was on 31 January 2016. Beyond being a regular visitor, we managed to receive invitations to press conferences. In this field we covered one whole event from the public relations view, namely the unforeseen birth of orangutan baby Rieke and the zoo's dealings with questions and possibilities regarding her situation from 12 January to 23 February 2015.[20]

Expedition Members

The expedition team consisted solely of Daniela Maria Hirsch (audio recordings, drawings and camera assistance) and David Gaehtgens (camera, audio recordings and AV-data handling). We specifically decided to be a small unit, to not be distracted from the situations at hand.

Both our backgrounds are in theatre production and filmmaking. Before the expedition, we previously worked on examining the nature of artistic documentation.

---

seeing well is far more important than smelling well, and the snout has shrunk considerably, giving the eye much better view'.

19   https://www.zoo-berlin.de/de/aktuelles/news/artikel/neuer-rekord [accessed 29 May 2017]

20   https://www.morgenpost.de/berlin/article206912589/Happy-Birthday-Rieke. html [accessed 29 May 2017]

Fieldwork & Research

We perceive the process from recording information to working with that information as a sensitive key activity. The seemingly neutral act of capturing information already contains a lot of shaping through parameters. Who is looking at what with which means? Medium as a neuter of *medius* simply denominates the middle. In natural science it describes the materials or the empty space through which signals, waves or forces pass. Artistically, medium refers to the material and to the technique being used for the production of artwork. This overlaps the definition of medium as a format for communicating and presenting information. In this regard we perceive the medium zoo as a message itself – a communication transmitted via a messenger.[21] We evaluated and selected our methodical approach in reference to this context. Due to the structure of time between the day trips, we could alternate intellectual research with fieldwork.

Methods

Looking At People Looking

A long time motive of our artistic interest is observing the observer. We had already developed this method in other works. Looking at the process of looking excludes that which is gazed at from the picture, but yet it is always present outside the framing as the clear direction of attention. Meanwhile the onlookers can be viewed both as an overall appearance and in the individuals' details that are beheld on each looker's side.

Direct Cinema

This is a form of documentary filmmaking. It developed simultaneously around 1960 in North America and in France by the filmmakers D.A. Pennebaker, the brothers Maysles and Jean Rouche. The aim was to produce documentation which is as genuine to the

---

21 Marshall McLuhan, *Understanding Media* (London: Routledge Classics, 2001), pp. 8-9: '[...] the content of any medium is always another medium. The content of writing is speech, just as the written word is the content of print, and print is the content of the telegraph. [...] This fact merely underlines the point that 'the medium is the message' because it is the medium that shapes and controls the scale and form of human association and action'.

objectified situation as possible. The approach tries to directly capture reality, unlike the mainstream media practice which heavily relies on *mise on scène* procedures. Key attributes are losing the commentary, working with lightweight cameras and equipment, to be a fly on the wall in a sense that there is no influence on what happens in front of the camera.[22] Yet, the presence of the filmmaker is always apparent. In the zoo all of our work was an embedded production, clearly visible, completely relying on the facilities at hand.

Long-term Recording

We did long-term video recordings, ranging from approximately twenty minutes to over two hours. Most of the situational accounting was done with one camera but sometimes we used two simultaneously. This captured two windows of observation onto one occurrence. We deliberately took our time to grow into our questions and so our working agenda shaped itself through the working process.

In addition to the technical chronicles, we also produced personal recordings with our own organic sensory equipment. This was also influenced by the long stretches of time we spent quietly in one place at a time, standing still in the movement of groupings. We produced layers of subjective memories in regard to our different personal interests. These recollections resulted in notes and drawings, both with no specific aim other than illustrating our personal experiences.

Participant Observation

This method of data collection aims at results through involvement with the observed situation. It originated in projects of fieldwork in the first half of the twentieth century. It influenced approaches in anthropology, enquiring about the qualitative aspects in field research. Bronisław Malinowski as a social anthropologist was one of the pioneering developers together with his students. His focus on the patterns of exchange stirred the review of personal contributions which are included in every process of collection.[23] The idea for this kind of

---

22   Compare *Primary*, dir. by Robert Drew (Drew Associates, 1960), 16mm

23   Bronislaw Malinowski, *Argonauts of the Western Pacific*, (London: Taylor&Francis e-Library, 2005), Preface by Sir James G. Frazer, p. V: ' Dr. Malinowski lived as a native among the natives for many months together,

ethnographic research was very different to the methods common in his time, which were more in a colonial tradition of information gathering.[24] The original standard in the work with curiosities of otherness by foreign cultures was to conduct interviews with an authoritarian approach, regarding the culture of interest as no equal partner. In contrast to that, participating in a cultural environment and even cultivating social relationships can enable a deeper and more open learning environment. There is a striking parallel in the development of animal and wildlife observation: the early TV programmes of David Attenborough follow the tradition of the top view. Just like with human animals, the study objects were caught and then observed and brought home. Today it is interesting to hear Attenborough talk about this time and how his views on documenting wildlife have developed and changed.[25]

In the zoo, we were clearly visible in and with our action. We had the same visitor status as the other visitors and also the same intentions in taking pictures. This is a common activity in the zoo. In fact, for us filmmakers it was a very special place in regard of camera freedom. In the zoo there is a consensual agreement that filming and photography is integral part of this zone. Very much in contrast to the camera consciousness and media sensitivity we now often have to deal with while working in shared public spaces. Through long-term recordings, we had another level of participation: taking our own subjective notes on the situation, often being better able to read it after a certain time of observation.

Results

Visual Compendium

All of the gathered material compiles into a body of visual knowledge. In the course of our field trips, we noticed that our camera results fall into three chapters: *Looking In*, *Looking Out* and *Safari*.

---

watching them daily at work and at play, conversing with them in their own tongue, and deriving all his information from the surest sources'.

24   e.g.: Hugo Bernatzik, *Gari Gari – Der Ruf der afrikanischen Wildnis*, (Vienna: L.W.Seidel&Sohn, 1938), p. 108: 'Ich lasse fürs erste von meinen Leuten den Elefanten ‚besetzen' und teile den Nuern mit, sie sollten sich sich solange gedulden, bis uns die Lage des Schiffes ermögliche, das Zerlegen des Tieres zu beobachten. Darüber gibt es eine erregte Auseinandersetzung mit den fleischwitternden Negern'.

25   *Zoo Quest in Colour*, 00:14:20, BBC 2017

These containers refer to the viewing angle in the framing. *Looking In* and *Looking Out* present a perspective into or out of a building. *Safari* relates on the one hand to the park setting of the zoo, on the other hand, it describes the peek, the wait and hunt for the view. Since the term derives from Arabic 'sāfara' where it translates as travel or journey, it also enquires about the quality of the movement itself and the nature of the little journey at hand while visiting the zoo.

Diorama

The Diorama originally presented a two-phase projection of images. Invented by Louis Daguerre and Charles Marie Bouton in 1822, it really was a picture-seeing device just as it literally means that which is seen.[26] The Diorama-experience soon took place inside purposely built and technically equipped buildings. They emulated theatres and included a revolving ground floor with the audience moving on it. In contrast to the much older Panorama, which is a wide-angled view often reaching up to 360°, the original Diorama is expanding into motion effects.[27]

However, popular understanding of the term later referred to something else than a building with a picture show installation. In the late nineteenth century, the newly emerged museums of natural history took to a version of the Diorama, in the aim of modelling settings of the natural world for permanent display.[28] These are a mix of two- and three-dimensional elements. The whole setting is equal in size to the scene it represents, the background being a painting and the foreground a lifelike sculptural replication of the original situation. This included taxidermy specimens, guaranteeing an unhindered view of the entire show case at all times.

---

26   Douglas Harper, *Online Etymology Dictionary,* 2001-2017, accessed 30 May 2017 at http://www.etymonline.com/index.php?allowed_in_frame=0&search=diorama

27   Sue Dale Tunnicliffe and Annette Scheersoi, ed., *Natural History Dioramas: History, Construction and Educational Role* (Dordrecht: Springer Science + Business Media, 2015), Chapter 2: Rainer Hutterer and Claudia Kamcke, *History of Dioramas*, p. 10: 'The diorama was originally focused on the representation of movements, because their absence in panoramas was felt as a deficiency'.

28   Ibid, p. 7: 'the meaning of the term 'diorama' changed through time and is still not clearly defined'.

Our visual material from the zoo is placed in between the two types of Diorama. Like the still models, we have the effect of the camera framing, which is similar to the display in a case. Inside this frame we have an animated showroom. Because in our material we portray the zoo's settings in a series of parts, our framings also suggest to be of prototypical explanation. The question of absent properties which might be of elementary influence on the situation at hand arises. This is something we aimed at intentionally. Furthermore, there is the reference to the general technical setup needed for the zoo, in many ways equalling the efforts of the immersive spatial installation of a Diorama building.

Tableau Vivant

The format of the living picture developed as a three-dimensional still life, an arranged presentation by actors which initially featured imitations of sculpted or painted statues. From the beginnings on theatre stages in the eighteenth century, it came into fashion for entertainment as well as educational purposes in the nineteenth century, when it also ventured into outdoor situations and into a broad portfolio of depicted motifs.[29]

While a central feature of the tableau is the art of keeping still, we perceived framings not without movement but with reduced activity. This is due to the activity range. In contrast to the intentional choice of actors in a tableau vivant to remain motionless, the capacity for acting on matters presented in the zoo's environment is slim. Therefore some parts of our material appear to be semi-motionless. This parallels the permanently moderate situation inside the models of habitat. It is especially linked to areas without majorly popular residents and hence reduced flow of audience. Also it occurs in the movement of the visitors themselves. Here it seems to be connected to the pursuit of gazing, reducing many other possible types of expression.

---

29   Brigitte Peucker, *The Material Image: Art and the Real in Film* (Stanford: Stanford University Press, 2007), p. 30: 'Tableau vivant is a meeting point of several modes of representation, constituting a palimpsest [...] overlay simultaneously evocative of painting, drama and sculpture'.

Discussions & Conclusions

In our results we noticed that the exhibition setup holds specific parameters. Central to its nature is, as described earlier, the allocation of roles. In the word 'role', deriving from French *rôle* as part or character one takes, we still have the roll of paper on which an actors part is written.[30] Today we also have the role model. This idea was promoted by the American sociologist Robert King Merton from 1957 onwards.[31] It could easily be embraced by the human frame of mind, because it is a concept the human species has been using for a long time. The distribution of designated positions is and has been the instrument of our human choice to serve as an explanatory structure in the complex universe we find ourselves in. We can compare the mythmaking in cosmological times to the reference group analysis in our anthropocentric times and find that the role is a basic dynamic in both: as a set of rights, obligations and expected behaviour patterns which are associated with a particular social status.

In the performing arts, these properties are openly scripted as a preset to the performance. They serve as an agreement on a particular plot and anchor certain qualities of commitment to a cause-and-effect relationship. Performance implies the act of constructing, producing and bringing about. Since exhibitions are public events, they imply and reflect social, economic and political developments. This is happening in real time and in a spacially manifested situation. Regarding this context, we determined that further enquiries are necessary into other formats of exhibition culture. Currently we are planning such follow up work.

The roles we recognize as central to exhibiting are:
– the Specialist = the establisher of fact-based knowledge
– the Object = the witness to the fact(s)
– the Visitor = the audience following the narrative at hand

In the situations we witnessed at the zoo, the basic characteristics of the visitor-performers were obviously clear to the audience. The actions

---

30   Harper, *Online Etymology Dictionary*, accessed 30 May 2017 at http://www. etymonline.com/index.php?term=role&allowed_in_frame=0

31   Robert King Merton, 'Social Structure and Anomie', *American Sociological Review*, Vol. 3, No. 5, October 1938, p. 674: 'The distribution of statuses and roles through competition must be so organized that positive incentives for conformity to roles and adherence to status obligations are provided for every position within the distributive order'.

we see on film are carried out in a calm and self-assured manner. This leads to the phenomenon that in the review of our material, all members of the audience often appear as if casted to participate in a commonly agreed production. And in fact, the image of the zoo is accessible in societies all over the globe. So there is a shared knowledge of the zoo's specific setting within human culture. Beyond that, the principle of the spectacle as a specially prepared or arranged display is widely in use and therefore familiar knowledge.

The part of the visitor does not hold many options for physical activity. The display is laid out for gazing, which involves mainly the eyes and the mind. Our fieldwork confirmed that the parameters to and within this exercise are of special importance. What is accomplished here?

Adventurous Activities

We attended most of the activities on offer. We expected the adventure of openly filming strangers. With experiences from other recording situations in contemporary public space, we were prepared for objection and protest. None occurred ever, not even when filming in the pets' corner where a high number of children were present. The only contact we had in connection to filming was regarding enquiries about our technical equipment, coming from fellow photographers.

No activities at the zoo demanded from us to take a brave chance, but the catering by the zoo's facilities.

Administration & Logistics

Due to our professional backgrounds, the tools we use for administering the topic at hand are all related to the discipline of aesthetics. The theoretical and practical study of receiving and processing sensory information to us seems a pragmatic starting point to inquire specific setups with a tentatively neutral stance. We look at a *phaìnomenon* in the etymological sense of 'that which appears'.[32] How does an appearance come forth through the parameters of human data-processing?

---

32   Harper, *Online Etymology Dictionary*, accessed 30 May 2017 at http://www. etymonline.com/index.php?term=phenomenon&allowed_in_frame=0

The action of gazing is intertwined with perception. This in itself is a product of a complex and synergizing system. Resulting from it are recognitions, bits of subjectively validated information. In the individual human mind, these bits assemble into understanding. But before interpretation lies the absorbing of information. Here, the exercise of attention does have a large impact. The position and quality of focalization does affect the conceived information. Attention always requires concentration, but the degree and extend of concentration can vary. Also predispositions due to personal experiences interact individually within the exercise of attention.

It is already becoming clear, that the 'coming into view' is an affair of multi-layered logistics. The starting point, however, is the human optical equipment which captures samples of sight. This visual survey is concerned with physical shape and form. The reading of outline and appearance is followed by an instant query on semblance and likeness. These comparative strategies for interpretation are involved even before further evaluation within the conscious mind. Basic referencing is a necessary tool for quick orientation, something of elemental value to all animals in regard to survival.

Therefore, the human civilization project has developed the use of images as readable representatives of meanings. The *imago* also reflects a likeness, it is imitating an appearance and therefore invites recognition. At the same time, it is transcending imitation, because it holds modified or specifically manufactured elements. After inviting the gaze into an accessible visual occurrence, imagery involves the following interpretation with content stretching beyond the initial visual impulse. It is a carrier of factual information but also of attitude and sentiment. The image is capable of transferring more than words and linear logic. This medium has been of great appeal to humankind throughout the varying stages of our species' development.

The method of visual communication was further empowered by techniques for mass-reproduction. The amount of images in everyday life increased due to mechanical publishing. The visual quality of pictures was influenced by available new technology, fusing word with image and this especially by the expansion of advertising. Also 'visualization' became a term describing, in the nineteenth century, the graphical display of data as precise mechanical drawings. The diagram is an exceptional achievement of abstraction in the human generation of visual media. Also it is a powerful component within organized presentation in regard to the establishment of coherent knowledge.

Conclusion

All animals use the experience of likeness for general orientation. Humankind also emphasizes otherness, addressing distinction and establishing a dualistic dynamic. This concerns human orientation in regard to affiliation and belonging.

The cultural practise of exhibiting is based on this dynamic. It draws on likeness for recognition and involvement, therefore enabling legibility. At the same time, the principle of display draws on otherness, because it creates a separation within the encounter. This separation stems from the properties of the allocated positions. I am the looker facing that which is looked at, which was purposefully chosen for display. All three positions hold different particularities, empowering certain qualities and diminishing others. For example: while the distance between the fixed positions can foster the awareness of perspective, the static setup itself increases the impression of stability and reliability. This is no purely intellectual experience. The exhibition is a medium that generates situational forms of knowledge. The way in which something is placed in relation to its surroundings constitutes an immediate quality and a departure point for all further effects. This is why the consideration of the gestalt as an overall shape and integral arrangement does matter greatly. By looking at the single parts, the functioning of the greater unit as a reciprocal structure cannot be seized.

A relevant dimension regarding exhibition culture emerges on this level. It is the dimension of the exhibit as evidence, resembling the meaning of this term in court. This is relevant because it asks about ramifications for the production of knowledge. Are we experiencing open or closed knowledge systems?[33] What are the systems' respective outcomes? And how important is this when the

---

33   Artist Luis Camnitzer's conception of 'art thinking' describes the museum as a system for organizing and acquiring knowledge. In his keynote lecture *Where is the Genie?* at the conference *The Idea of the Global Museum*, December 2-3 2016 at Hamburger Bahnhof – Museum für Gegenwart Berlin, he refers to the dynamics in closed and open knowledge systems. He identifies as central the autodidactic processes within the viewer: a closed knowledge system is asking 'Do you get it?' and an open approach is asking 'What will you do with it?'. In the next step, the closed system asking 'Do you like it?', the open question is 'How will you carry it on your own?'.

use of designated sights has become a general standard within a more and more global human culture?[34]

Another holistic dynamic is the plot, the overall direction a context is driving towards. At the zoo this is a human passage from nature to culture. As in every narrative, certain signals which are shared inside a group of recipients serve as a familiar basis of anticipation.[35] These signals draw from habits and conventions, they confirm and consolidate meaning. They also support judgement and participate in the creation of symbols, our human tokens of joint remembrance. Here, there is room for projection, the transference of desires, feelings or intentions onto beings, objects and circumstances. In fifteenth century alchemy, projection describes the transmutation of a lesser substance into a higher form. Today we can witness the effect of projecting additional content into a given constellation, transforming or disguising the initial experience.

The exhibition format has become general situational knowledge. The rules of engagement with this format are familiar throughout human societies. Though not all parts of a society become physically involved with the rooms of exhibition spaces, they are accepted as a vehicle of human culture. This culture rather meets opposition and critical thinking from individuals actively working within this space, but never fundamentally questioned from the outside visitors. Disengagement or detachment are the silent expressions of dislike. As mentioned before, we actively pursue further questioning regarding these interrelations.

A specific characteristic of the zoo is the required level of previous engagement: because the subject of the exhibition seems to be the natural world, every visitor can easily bring some expertise. The most basic expertise is being an animal ourselves, even if this is not consciously perceived. It is a subject of low-threshold, balancing possibly educative ambitions with light entertainment. In this special atmosphere one striking property of the space almost escapes apperception – it is the

---

34　Guy Debord, *Die Gesellschaft des Spektakels*, trans. by Jean-Jacques Raspaud (Luzern: Edition Libertaire 1994), No. 5: 'Das Spektakel kann nicht als Übertreibung einer Welt des Schauens, als Produkt der Techniken der Massenverarbeitung von Bildern begriffen werden. Es ist vielmehr eine tatsächlich gewordene, ins Materielle übertragene Weltanschauung. Es ist eine Anschauung der Welt, die sich vergegenständlicht hat'.

35　Norbert Elias, *Über den Prozess der Zivilisation*, 8th edn. (Frankfurt am Main: Suhrkamp Verlag, 1981), p. XLVI: '[...] Verständnis langfristiger Prozesse, die Menschen auf der individuellen und auf der gesellschaftlichen Ebene gleichzeitig durchlaufen'.

artificial nature of the seemingly naturalistic display.[36] The scenery is entirely fabricated, not even falling back onto the original matter of the referenced situation. Instead a proxy is created, recreating outline and guise from different material than the original it represents. A solid stone becomes a man-made object, with a plastered and painted surface. The familiar setting of the park frames the naturalistic replica. The enclosed garden is a popular man-made naturalistic scene. The use as a zoological garden makes it difficult to decode the constellation of *Ersatz*, the artificial replacement differing in kind from and inferior in quality to what it replaces. In the zoo we are dealing, on many levels, with the concept of the substitute as something used in place of something else – gradually, we realized, this captures the transitive nature of the civilization process itself. This realization transformed our perception of the zoo into an allegory of the human civilization project, speaking about its quality and nature. And, as this is an ongoing process, so was and is the zoo exhibition practice itself in constant evolvement.

Our framings are windows for observation and as in every allegory an important move is from 'What is it?' to 'What does it do?'. With the playback in installative situations we now want to find out what kind of potential the material has in this regard. A continuous re-search: to look and look again and look again.

Finally, we would like to add a personal conclusion on the zoo: to us it is also a meditation on the human culture of caging. This practise originated within humankind and is not shared by any other species. As an activity it has shaped human development on many levels. The zoo is an opportunity to register the texture and character of this pursuit of barring. At the zoo, we wish for the 'looking at' to be 'listening in'. And after listening, so we dream, talk and interchange on this complex could be activated on all channels of human sensitivity, challenging the culture of gazing into motion.

Acknowledgements

We would like to thank Barbara and Thomas Gaehtgens for their generous financial support, which was greatly enabling this expedition. Further we would like to thank all open minds past and present, that

---

36   Baratay, p. 90: 'Der Landschaftsgarten verfährt nämlich nach der Methode des Fragmentierens und Collagierens'.

have contributed thoughts and feedback, keeping us alert and agile during this expedition. Mostly we want to thank the resident animals at the zoo for their incredible expense.

Bibliography

Allin, Michael, *Zarafa*, trans. by Wolfgang Schuler (Munich: Diana Verlag, 1999)

Attenborough, David (Director), *Zoo Quest (BBC, 1954-1964)*

Beer, Bettina, ed., *Methoden ethnologischer Feldforschung*, 2nd edn. (Berlin: Dietrich Reimer Verlag, 2008)

Berger, John, *Why Look at Animals?* (London: Penguin Books, 2009)

Bernatzik, Hugo, *Im Reich der Bidjogo – Geheimnisvolle Inseln Afrikas*, (Berlin: Deutsche Buchgemeinschaft, 1933)

Bernatzik, Hugo, *Gari Gari – Der Ruf der afrikanischen Wildnis*, (Vienna: L.W.Seidel & Sohn, 1938)

Blunt, Wilfrid, *The Ark in the Park – The Zoo in the Nineteenth Century* (London: Hamish Hamilton Ltd, 1976)

Bredekamp, Horst, *Antikensehnsucht und Maschinenglauben* (Berlin: Verlag Klaus Wagenbach, 2000)

Drew, Robert (Director), *Primary* (Drew Associates, 1960), 16mm

Debord, Guy, *Die Gesellschaft des Spektakels*, trans. by Jean-Jacques Raspaud (Luzern: Edition Libertaire 1994)

Elias, Norbert, *Über den Prozess der Zivilisation*, 8th edn. (Frankfurt am Main: Suhrkamp Verlag, 1981)

Eric Baratay, Eric, and Hardouin-Fugier, Elisabeth: *Zoo – Von der Menagerie zum Tierpark*, trans. by Matthias Wolf (Berlin: Verlag Klaus Wagenbach, 2000)

Geertz, Clifford, *Dichte Beschreibung – Beiträge zum Verstehen kultureller Systeme*, trans. by Brigitte Luchesi and Rolf Bindemann (Frankfurt am Main: Suhrkamp Verlag, 1983)

Hahn, Hans Peter, *Ethnologie* (Berlin: Suhrkamp Verlag, 2013)

Hahn, Hans Peter, *Materielle Kultur*, 2nd edn. (Berlin: Reimer Verlag, 2014)

Heck, Lutz, *Aus der Wildnis in den Zoo* (Berlin: Verlag Ullstein, 1930)

Holert, Tom, ed., *Imagineering* (Cologne: Oktagon Verlag, 2000)

Humboldt, Alexander von, *Ansichten der Natur* (Stuttgart: Reclam Verlag 1969)

Larson, Greger, and Fuller, Dorian: 'The Evolution of Animal Domestication', *Annual Review of Ecology, Evolution, and Systematics,* Vol. 45, 2014

Lévi-Strauss, Claude, *Das wilde Denken*, 16th ed., trans. by Hans Naumann (Berlin: Suhrkamp Verlag 2013)

Lorenz, Konrad, *Das Jahr der Graugans*, 6th edn. (Munich: Deutscher Taschenbuch Verlag,1990)

Malinowski, Bronislaw, *A Diary in the Strict Sense of the Term* (London: Routledge and Kegan Paul Ltd, 1976)

Malinowski, Bronislaw, *Argonauts of the Western Pacific*, (London: Taylor&Francis e-Library, 2005)

Merton,Robert King: 'Social Structure and Anomie', *American Sociological Review*, Vol. 3, No. 5, October 1938

McLuhan, Marshall, *Understanding Media* (London: Routledge Classics, 2001)

Morris, Desmond, *The Naked Ape* (Frogmore: Triad / Mayflower Books, 1977)

Peucker, Brigitte, *The Material Image: Art and the Real in Film* (Stanford: Stanford University Press, 2007)

Schillings, C.G., *Mit Blitzlicht und Büchse* (Leipzig: R.Voigtländers Verlag, 1922)

Schmitz, Friederike, ed., *Tierethik*, (Berlin: Suhrkamp Verlag 2014)

Tunnicliffe, Sue Dale, and Scheersoi, Annette, ed., *Natural History Dioramas: History, Construction and Educational Role* (Dordrecht: Springer Science + Business Media, 2015)

Ullrich, Jessica, ed., *Tierstudien 07: Zoo* (Berlin: Neofelis Verlag, 2015)

Virilio, Paul, *Der negative Horizont*, trans. by Brigitte Weidmann (Munich: Carl Hanser Verlag, 1989)

Winser, Shane, ed., Royal Geographical Society Expedition Handbook (London: Profile Books, 2004)

AUTHORS

MARIAELISA DIMINO is a PhD student in German Literature at the University of Verona. She is currently working on a dissertation on Alfred Kubin and the intersections between literature and visuality. She was member of the organizing and scientific committee of the international conference 'Bestiarium. Human and Animal representations', held at the University of Verona (28-30 September 2016). In September 2015 she took part in the international conference 'Robert Musil als Redakteur der Tiroler Soldatenzeitung' with a paper entitled 'Propaganda, Werbung und die Darstellung der Kriegserfahrung: Robert Musils Umgestaltung der Tiroler Soldatenzeitung'. Among her recent publications are 'Il duplice demone di Alfred Kubin', in Giacomo Raccis (ed.), Elephant&Castle n.15, *Scrivere, vedere, dipingere prospettive transmediali per lo Studio della letteratura* (2016), 'Der Totentanz im elektronischen Unbewussten: Die Videokunst von Alessandro Amaducci', in Nitsche J. (ed.), *Mit dem Tod Tanzen. Tod und Totentanz im Film*, Neofelis Verlag, Berlin, 2015; John Webster, *Il Diavolo Bianco* (traduzione, note e saggio introduttivo), Catania, Villaggio Maori Edizioni, 2014.

ALESSIA POLATTI graduated in Language, Society and Communication at the University of Bologna and is now a PhD candidate in English Postcolonial Literature at the University of Verona. Her PhD project deals with the experiences of migration and diaspora of Black British authors, with a focus on the phenomena of 'return' and 'reverse' migration. Among her research interests are Black British Fiction, 19th century colonial literature, the postcolonial rewriting of the English canon, and the interconnections between Postcolonial Literature and globalization. She was member of the organizing and scientific committee of the international conference 'Bestiarium. Human and Animal representations', held at the University of Verona (28-30 September 2016). In November 2016 she participated in the Second Callaloo Early Career Researchers Workshop *'Race' and the Academy since 1800*, at TORCH -University of Oxford with a paper on Rider Haggard and Andrew Lang's 'academic' friendship between racial theories and colonial romances. Among her recent publications is 'A Struggle between Literary and Self-Cannibalization: The Brontës' Reversal in V. S. Naipaul's *Guerrillas*', in *Il Tolomeo* (2016).

ROBERTA ZANONI is a PhD candidate in Foreign Languages, Literatures and Cultures at the University of Verona. Her research concerns the use made by contemporary advertising of the works and figure of Shakespeare. Her main interests are English Literature, Literature and Mass Media, and Literary Transpositions and Translations. She was member of the organizing and scientific committee of the international conference 'Bestiarium. Human and Animal representations', held at the University of Verona (28-30 September 2016). She has recently published an essay 'La manipolazione del linguaggio in *Riccardo III*' in *Riccardo III dal testo alla scena* edited by Mariangela Tempera, a review of *Literature and Human Rights* by Ian Ward and a review of *Discoursive Framings of Human Rights*, by Karen-Margrethe Simonsen and Jonas Ross Kjærgård. She participated in the World Shakespeare Congress with a paper on 'Shakespeare in Pop Cinema' and in the Britgrad 2017 Conference, held at Stratford-Upon-Avon, with a paper on 'Political Discourse in Richard III'.

LUCIA ZAIETTA is Honorary Fellow at the University of Pavia (Italy). She obtained a Ph.D. in Philosophy (November 2017) at the Northwest Italian Philosophy Consortium (FINO) and University of Paris Panthéon Sorbonne, under the supervision of Renaud Barbaras and Luca Vanzago. Her dissertation concerned the notion of animality in relation to Merleau-Ponty's phenomenology. She attended several academic conferences and published articles in Italian, English and

French. Her main research interests are phenomenology and animal studies, with particular attention to the subjects of nature and life.

RICHARD HUTCHINS is a PhD candidate in Classics and The Program in Classical Philosophy at Princeton University. His dissertation, entitled *Lucretius Against Human Exceptionalism*, explores human and animal superiority in Lucretius' *De rerum natura*. His research combines the Environmental Humanities with Classical Philosophy, and focuses on the Presocratics, Epicureanism, Seneca, and the Neoplatonist Porphyry. He is also interested in Hedonism from antiquity to the present. Richard is also a member of the Postclassicisms Network at Princeton University, and is the co-founder of Living Greek in Greece, a spoken ancient Greek program that takes place every summer in Selianitika, Greece.

AMADEUSZ JUST is a PhD candidate at the Institute of Philosophy, University of Warsaw. He studied ethnology and philosophy; he is currently working on a doctoral dissertation entitled "Non-discursive social practices. An attempt to surpass the rational paradigm of social philosophy". He is a member of Austrian Ludwig Wittgenstein Society.

BENEDETTA PIAZZESI is a PhD student in philosophy at Scuola Normale Superiore of Pisa. She has dealt with the history of animal husbandry in a Foucauldian perspective and is currently developing a project on the history of ethology, namely on the scientific acquisition of the problem of animal intelligence and behaviour. She has already published for Mimesis *Così perfetti e utili. Genealogia dello sfruttamento animale* (2015) and *Un incontro mancato. Sul fotoreportage animalista* (2017). For the publisher Raffaello Cortina, in collaboration with Gianfranco Mormino and Raffaella Colombo, she has published *Dalla predazione al dominio. La guerra contro gli animali* (2017). She is part of the editorial staff of *Liberazioni. Rivista di critica antispecista*.

PETER KOFLER is Associate Professor of German Literature at the University of Verona. He particularly deals with the work of Christoph Martin Wieland, the intercultural relationship between Italy and

Germany, literature and visual arts, literature and music, and the history and theory of literary translation. Among his publications are the edition of the translation of Dante's *Inferno* by Christian Joseph Jagemann, as well as essays on Gotthold Ephraim Lessing, Wilhelm Heinse, Johann Wolfgang Goethe, Theodor Fontane, Hugo von Hofmannsthal, Robert Musil, Franz Kafka, and Paul Celan.

FLAVIA PALMA obtained a BA in Italian Literature (2011, 110/110 *cum laude*) and a MA in Tradition and Interpretation of Literary Texts (2013, 110/110 *cum laude*) from the University of Verona. In 2017, she was awarded a doctorate in Philology, Literature, and Linguistics from the University of Verona, receiving also the certificate of *Doctor Europaeus*. In her doctoral thesis she analysed the functions of paratexts and framing devices in Italian collections of novellas and in English collections of tales influenced by the Italian novella tradition, during a period ranging from the mid-thirteenth to the early-seventeenth centuries. She participated in several international conferences in Italy and abroad, and spent various research periods in the UK. She published articles devoted to the reception of Italian novellas in England, to the employment of proper names and pseudonyms in Renaissance novellas' frame tales, and to Girolamo Parabosco's literary and dramatic works. She is currently a post-doctoral researcher in Italian Literature at the University of Verona.

SIMONE REBORA holds a PhD in Foreign Literatures and Literary Studies (University of Verona) and a BSc in Electronic Engineering (Polytechnic University of Torino). Between 2016 and 2017, he held a DAAD fellowship at the University of Göttingen. Currently, he works as a research fellow between the University of Verona and the Göttingen Centre for Digital Humanities. He is a member of the International Society for the Empirical Study of Literature and Media (IGEL) and of the Evolution of Reading in the Age of Digitisation (E-READ) European COST Action. His main research interests are theory and history of literary historiography and reader response studies. In the field of digital humanities, he focused on tools and methods like OCR, stylometry, and sentiment analysis. His essays have appeared in journals such as *Antologia Vieusseux, Between,* and *Modern Language Notes.* In 2015, he published the monograph *Claudio Magris.*

EIRINI APANOMERITAKI is a PhD Candidate in Comparative Literature at the University of Essex. Her thesis examines the narrative representations of transforming subjectivity across the twentieth century, with special focus on literary texts that deploy the trope of metamorphosis. Her doctoral project encompasses selected short stories and novels from different literatures, from 1917 to 1999, including the fiction of Franz Kafka, Vladimir Nabokov, Virginia Woolf and Marie Darrieussecq. Her research interests focus on theories of subjectivity and identity in modernist and post-modern literature, psychoanalysis, gender and sexuality and the reception of myth in modern and contemporary literature and film. She serves on the executive committee of the Centre for Myth Studies. She is currently teaching psychoanalytic theory and popular culture in the Department of Psychosocial and Psychoanalytic Studies at the University of Essex.

MARIE CAZABAN-MAZEROLLES is a last-year PhD student in Comparative Literature at the University of Poitiers in France. Her research focuses on the way post-darwinian life sciences impacts on narrative poetics among 20th and 21st centuries fictions.

VALENTINA SAVIETTO studied German and Romance languages and literatures at the University of Padua and Würzburg and graduated with honours with a thesis about Thomas Mann's novel *Doktor Faustus* (2011). She did a doctorate at the University of Verona, where she was awarded a PhD in German Studies in April 2017, through a co-tutoring agreement with OttoFriedrich-University of Bamberg. Her doctoral thesis focuses on Klaus Mann's fictional works and analyses them from the point of view of Interart and Intermedial Studies. Her research interests range from Intermedial Studies and the Manns to the German Romanticism (E.T.A. Hoffmann), the genre of musical novel, the reception of Nietzsche and Wagner in literary texts and also to Paul Heyse. She now works as lecturer at the University of Verona. Previously, she attended a Music Conservatory in Castelfranco V.to (TV) and got a Viola-degree in 2010.

JUDITH RAHN studied English and German Literature at Rheinische-Friedrich-Wilhelms-Universität Bonn and the University of Oxford. She is currently working on her Ph.D. project with the working title 'Exploring Posthuman Life in Contemporary Fiction' and holds a research position and lectureship at the Department of English and American Studies at Heinrich-Heine-University Düsseldorf. Her research interests include posthuman theory, black British fiction, new materialism, anthropophagy, and the fantastik.

ELENA OGLIARI holds an MA in English and Russian Literature from the State University of Milan, where she graduated with a thesis on the literary relationship between Henry James and Ivan Turgenev. She is currently a PhD candidate in the Department of British Culture at the same University. Among her publications on James's oeuvre and personality: 'Beyond the Historical Record? Henry James in 'The Master at St. Bartholomew's Hospital 1914–1916'', which analyses a short-story by Joyce Carol Oates in the context of the *Jamesian biofiction*, 'Henry James Goes On A Diet: A Chronicle Of A Private Drama' and the forthcoming 'A Second Thought on Henry James's First Professional Letters'.

DANIELA MARIA HIRSCH (*1973, University Burg Giebichenstein Halle/Saale) and DAVID GAEHTGENS (*1973, Academy of Media Arts Cologne) are collaborating as gaehtgens.hirsch since 2012. Backgrounds in photography, film-making, theatre-production and media-art are at the base for artistic research on the human condition. The dynamics of fabricating images are of special concern in their work, because the production of images is unique to humans. With a view on the evolution of media, they experiment with formats for communicating and presenting information. Their starting point has always a documentary nature. They embed pictures in installative environments and ask about the nature of situations. What are the performative aspects of static artwork while on display? Which dynamics of roles and expectations come together in the exhibiting situation? And what purpose and function is connected to exhibition culture? Currently they are researching objects in the moving Ethnological Museum of Berlin.

INDEX OF NAMES

Printed by Agrisys Holding SA - May 2018